D0207395

# Introduction to molecular symmetry

## J. S. Ogden

University of Southampton

Series sponsor: AstraZeneca

AstraZeneca is one of the world's leading pharmaceutical companies with a strong research base. Its skill and innovative ideas in organic chemistry and bioscience create products designed to fight disease in seven key therapeutic areas: cancer, cardiovascular, central nervous system, gastrointestinal, infection, pain control, and respiratory.

AstraZeneca was formed through the merger of Astra AB of Sweden and Zeneca Group PLC of the UK. The company is headquartered in the UK with over 50,000 employees worldwide. R&D centres of excellence are in Sweden, the UK, and USA with R&D headquarters in Södertälje, Sweden.

AstraZeneca is committed to the support of education in chemistry and chemical engineering.

## OXFORD
### UNIVERSITY PRESS

BOWLING GREEN STATE
UNIVERSITY LIBRARIES

# OXFORD

UNIVERSITY PRESS

Great Clarendon Street, Oxford OX2 6DP

Oxford University Press is a department of the University of Oxford.
It furthers the University's objective of excellence in research, scholarship,
and education by publishing worldwide in

Oxford New York

Athens Auckland Bangkok Bogotá Buenos Aires Cape Town
Chennai Dar es Salaam Delhi Florence Hong Kong Istanbul Karachi
Kolkata Kuala Lumpur Madrid Melbourne Mexico City Mumbai
Nairobi Paris São Paulo Shanghai Singapore Taipei Tokyo Toronto Warsaw
with associated companies in Berlin Ibadan

Oxford is a registered trade mark of Oxford University Press
in the UK and in certain other countries

Published in the United States
by Oxford University Press Inc., New York

© J.S. Ogden 2001

The moral rights of the author have been asserted
Database right Oxford University Press (maker)

First published 2001

All rights reserved. No part of this publication may be reproduced,
stored in a retrieval system, or transmitted, in any form or by any means,
without the prior permission in writing of Oxford University Press,
or as expressly permitted by law, or under terms agreed with the appropriate
reprographics rights organizations. Enquiries concerning reproduction
outside the scope of the above should be sent to the Rights Department,
Oxford University Press, at the address above

You must not circulate this book in any other binding or cover
and you must impose this same condition on any acquirer

Library of Congress Cataloguing in Publication Data
(Data applied for)
ISBN 0–19–855910–0
1 3 5 7 9 10 8 6 4 2

Typeset by Newgen Imaging Systems (P) Ltd., Chennai, India

Printed in Great Britain by
The Bath Press, Bath

# Series Editor's Foreword

Oxford Chemistry Primers are designed to give a concise introduction to all chemistry students by providing the material that would usually form an 8–10 lecture course. As well as providing up-to-date information, this series expresses the explanations and rationales that form the framework of the current understanding of inorganic chemistry.

Symmetry is fundamental in science, architecture and art. Steve Ogden here provides an introductory course into symmetry and the chemical applications of group theory. This is an essential skill for tackling many bonding and spectroscopy problems. Here he concentrates on molecular vibrations and chemical bonding. We benefit from Steve's long experience both in teaching this subject and also in his own careful research identifying molecules formed under high temperature conditions using vibrational spectroscopy. Symmetry is a core aspect of modern undergraduate chemistry and this book provides an appropriate basis for understanding the elegance and simplifications attainable by its application.

John Evans
*Department of Chemistry*
*University of Southampton*

# Preface

One of the challenges faced by any lecturer delivering a course on molecular symmetry is to select and present material which will provide an adequate mathematical grounding in this area, whilst at the same time retaining strong links with "real" chemistry. Most students are familiar with the basic concepts of symmetry, and the initial aim of this Primer is to provide a suitable framework for describing the symmetry properties of molecules both pictorially, and by the use of matrices. For some students, this may be their first encounter with matrices, and the approach adopted here is to regard matrices simply as descriptive tools rather than to explore their intrinsic properties in any detail. These initial chapters are then followed by a description of the role played by symmetry in vibrational spectroscopy and in chemical bonding. Examples abound in both these areas of chemistry, and it has been necessary to be somewhat arbitrary in the choice of molecules studied. However, the examples included here cover many of the basic molecular structures which would be encountered in a typical university chemistry course.

This Primer would not have been written without the initial encouragement of John Evans, and I would like to give particular thanks to him for his constructive comments and keen eye for detail throughout the production of the draft manuscript. In the final stages of production, numerous Southampton students have also had the opportunity to comment on the material in this Primer, and to them also, I would like to express my appreciation. However, my main thanks must go to my family for their ability simultaneously both to support and cajole me into finally completing this short book.

*Southampton*                                                                                    J.S.O.
February 2001

# Contents

# Introduction—the structure of this Primer

Symmetry is all around us, and even from early childhood most people seem to have an innate ability to respond to those shapes and patterns which can be described in terms of symmetry. At a conscious level, it underpins our notions of 'fairness'—during cake-cutting operations, for example—and at a subconscious level it has even been claimed that the more symmetrical a person's face, the more attractive it is to the opposite sex.

And if further confirmation should be required of the influence symmetry may have on our lives, one need look no further than the advertising world. Here, real money has been spent trying to catch and hold our attention with a multitude of logos and emblems, many of which utilise designs with characteristically identifiable symmetry. This apparently innate response to symmetry, and the attendant search for pattern, surface in numerous ways in almost every branch of science, and descriptions or 'explanations' which incorporate symmetry are often regarded not only as being particularly elegant, but possibly even 'true'.

This Primer is concerned with the symmetry of molecular species, and with the inter-relationship between molecular symmetry and properties such as vibrational spectra and molecular energy levels.

In order to explore these aspects of chemistry, it is necessary first to establish a framework for describing in a consistent way both the symmetry of a particular molecule, and also the results of carrying out symmetry operations. The former task leads to the idea of the molecular point group, and occupies a major part of Chapter 1.

Chapter 2 begins with a description of single symmetry operations in terms of simple matrices, continues with illustrations of sequential symmetry operations, and leads to the concept of a symmetry group through a consideration of multiplication tables. This is followed by an introduction to representations and basis functions.

Character tables, and the use of the reduction formula to establish the symmetries of irreducible representations are described in Chapter 3, and other basis functions such as interatomic bonds and atomic orbitals are also introduced here, whilst degenerate representations and higher order point groups are explored in Chapter 4.

Chapter 5 deals with the symmetry aspects of vibrational spectroscopy, using firstly a fuller treatment which includes bending modes, and secondly a 'stretches-only' approach based on internal coordinates. In Chapter 6, the role of symmetry in infrared and Raman vibrational spectroscopy is discussed.

The final chapter provides an introduction to symmetry aspects of molecular bonding, and in particular to the sets of orbitals which can combine to form a molecular orbital diagram.

The book has been designed to be of assistance to Chemistry students at several different stages during their degree courses. An understanding of symmetry elements and operations, and the resulting description of molecular symmetry in terms of the point group, now form part of many First Year Honours courses; whilst in their Second and Third years, students increasingly encounter aspects of spectroscopy and bonding which greatly benefit from a symmetry-based approach involving the use of character tables.

Each chapter contains a number of worked examples, and there are also supplementary problems for individual or group study. These exercises are of varying degrees of difficulty, and although their primary purpose is to reinforce directly the material in the chapter, some of them, at least, may be of interest to graduate students with a weakness for problem-solving. The answers to these appear in Appendix I.

Appendix II contains the character tables of the most commonly encountered point groups, the understanding and use of which are central to many areas of modern chemistry, and a bibliography of additional reading may be found in Appendix III.

# 1 Symmetry elements, symmetry operations and point groups

From their earliest years at school, children learn to describe 'symmetrical' shapes in terms of two distinct symmetry elements—a plane, or 'line' of symmetry, and an axis of symmetry—and these two symmetry elements remain the most easily identifiable ones exhibited by the majority of 'symmetrical' molecular species. In general, a shape possesses a plane of symmetry if the operation of reflection in the plane results in an equivalent mirror image. In a similar way, a shape possesses an axis of symmetry when simple rotation about such an axis leads to an equivalent configuration.

## 1.1 Plane of symmetry: symbol $\sigma$

At this stage, it is convenient to introduce the scheme of notation used in this Primer to denote elements of symmetry. The terminology used throughout this book is due to Shönflies and, using this scheme, a plane is denoted in general by the symbol $\sigma$. In many instances, however, it is necessary to be more specific regarding the location of a plane within a molecule, and various subscripts such as h (for horizontal), v (vertical), d (dihedral) or $xz$ are added to identify the position of the plane either in relation to other symmetry elements or to an established coordinate system.

Figure 1.1 shows the position of the single symmetry plane in the pyramidal molecule $PCl_2F$. If the coordinate system chosen for this molecule is as shown, this would be the $\sigma(xy)$ plane.

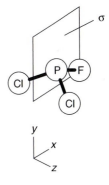

**Fig. 1.1**

## 1.2 Proper axis of rotation: symbol $C_n$

A simple or proper axis of rotation is denoted by the symbol $C_n$, where $n$ is termed the order of the axis. This symmetry element is present whenever rotation through an angle of $360°/n$ results in an equivalent configuration.

Figure 1.2 shows a variety of simple shapes which illustrate the presence of rotation axes $C_n$ ($n = 2$–$4$).

In addition to its significance as a symmetry element, the existence of a $C_n$ axis is important in defining the orientation of a shape in relation to everyday terminology such as 'vertical' or 'horizontal'. Basically, if a shape contains only one axis of symmetry, then the direction of this axis is identified as 'vertical', whatever orientation it may happen to appear in a textbook. If a shape contains axes of different orders, then 'vertical' is associated with the direction of the axis of highest order.

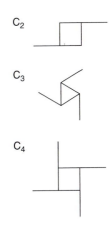

**Fig. 1.2**

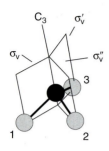

**Fig. 1.3** The shape of $PCl_3$.

**Fig. 1.4**

**Fig. 1.5**

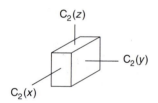

**Fig. 1.6**

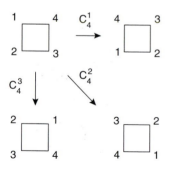

**Fig. 1.7** Operations of a $C_4$ axis.

Once 'vertical' has been established, it can be used to describe the nature of any planes present in a shape. Such planes might contain the vertical axis—and be denoted as $\sigma_v$ or $\sigma_d$—or they may be perpendicular to it—and hence 'horizontal' planes, $\sigma_h$. The distinction between $\sigma_v$ and $\sigma_d$ will be discussed later.

Figure 1.3 shows a sketch of a tilted $PCl_3$ molecule. In addition to the $C_3$ axis, there are three planes of symmetry, denoted as $\sigma_v$, $\sigma_v'$ and $\sigma_v''$ but despite appearances, all three of these planes would be described as vertical planes.

Occasionally, it happens that there is no single high order axis from which the terms 'vertical' and 'horizontal' may be uniquely derived.

Figures 1.4 and 1.5 illustrate the presence of three two-fold axes of rotation in a simple diamond shape, and in the capital letter 'H'. Two of the $C_2$ axes lie in the plane of the page and are identified here as $C_2$ and $C_2'$, and there is a third $C_2$ perpendicular to the plane. The rectangular prism shown in Fig. 1.6 also possesses three mutually perpendicular $C_2$ axes. For these shapes, the terms 'vertical' and 'horizontal' are somewhat arbitrary, and the situation is resolved by choosing the three axes to be the Cartesian coordinates $x$, $y$ and $z$, and denoting them as $C_2(x)$, $C_2(y)$ and $C_2(z)$ respectively. The planes of symmetry in these shapes would then be identified as $\sigma(xz)$, $\sigma(yz)$ and $\sigma(xy)$.

## 1.3   Elements and operations

In the diamond shape shown above, the presence of a $C_2$ axis or plane is effectively defined through its attendant symmetry operation—rotation or reflection—and it may seem unnecessary to make a distinction between the symmetry element and its operation. However, there is an important difference between these two terms, and this may be illustrated by considering the symmetry elements and operations for the $PCl_3$ structure shown in Fig. 1.3. Here, each of the three planes of symmetry is distinct, each having just one readily identifiable operation. The plane labelled $\sigma_v$ interchanges atoms 2 and 3, whilst leaving atom 1 unshifted, and the other two planes similarly have simple operations.

However, the single rotation axis, $C_3$, can generate *two distinct operations*, each of which results in a configuration equivalent to the starting position: anticlockwise rotation through 120° (1 to 2), and also rotation through 240° (1 to 3). These separate operations are denoted by the symbols $C_3^1$ and $C_3^2$ respectively, but there is only one $C_3$ axis.

In a similar way, one can visualise higher order axes, such as the $C_4$ axis present in a simple square. This gives rise to the three distinct operations $C_4^1$, $C_4^2$ and $C_4^3$, corresponding to rotation through 90°, 180° and 270° respectively, as shown in Fig. 1.7.

The final stage in the $C_n$ rotational sequences—rotation through 360°—would be denoted as $C_n^n$, but would produce a configuration *identical* to the starting position, not *merely equivalent*. It is therefore not regarded as an operation of the $C_n$ axis. As discussed below, a special symmetry element, the identity (symbol E), is equivalent to this operation.

## 1.4 Coincident axes

A closer examination of the operation $C_4^2$ in Fig. 1.7 shows that this operation corresponds to rotation through 180°, and is therefore identical to the effect of a two-fold rotation about the same axis. As a result, we conclude that there is a $C_2$ axis coincident with $C_4$. The effect of this on the symmetry operations of the $C_4$ axis is essentially to remove $C_4^2$ from the group of rotations associated with $C_4$ and to regard it instead as a $C_2$ operation.

A similar situation exists when considering other shapes with high order $C_n$ axes such as the regular hexagon. Here (Fig. 1.8), the five anticlockwise operations $C_6^1$, $C_6^2$, $C_6^3$, $C_6^4$ and $C_6^5$ will take an atom at position 1 to positions 2, 3, 4, 5 and 6 respectively. However, the operation $C_6^3$ is equivalent to rotation through 180°, whilst $C_6^2$ and $C_6^4$ correspond to rotation through 120° and 240° respectively. We can therefore identify *both a* $C_2$ *and a* $C_3$ *which are coincident with the* $C_6$. The operation $C_6^6$ is similarly described by the symbol E.

The net effect of this is that the proper rotations associated with this axis are classified as $C_6^1$, $C_3^1$, $C_2$, $C_3^2$ and $C_6^5$.

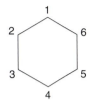

**Fig. 1.8**

## 1.5 Rotation–reflection axis: symbol $S_n$

In order to represent the rotational equivalence exhibited by some shapes, it is necessary to introduce a second type of rotational symmetry axis—the rotation–reflection axis. This has the general symbol $S_n$ and the operations associated with this axis are illustrated below. This symmetry element is also known as an improper axis of rotation.

Figure 1.9(a) depicts the relative atom positions in $B_2Cl_4$. This molecule contains one B–B bond (joining atoms 5 and 6) and four equivalent bonds from the boron atoms to Cl atoms at positions 1–4. The structure is perhaps best visualised in relation to a square-based prism, as shown. There is a $C_2$ axis along the B–B bond, which relates the Cl atoms at positions 1 and 2 to 3 and 4 respectively. However, *all four* Cl atom positions are related by a *new* symmetry element capable of generating, for an atom at position 1, the sequential transformations 1–2–3–4–1. The symmetry element which can achieve this is the rotation–reflection axis $S_4$, *which also lies along the B–B bond.*

As its name suggests, this $S_4$ axis has a two-stage mode of operation. The first stage is *rotation through 90°*, and corresponds to what would be expected for the $C_4^1$ operation, taking each atom to the intermediate positions shown in Fig. 1.9(b). This rotation is then followed by a *reflection* in a plane perpendicular to the B–B axis (Fig. 1.9(c)). The boron atoms 5 and 6 would also be affected by this second step.

It must be emphasised that this perpendicular plane need not be an actual plane of symmetry present in the molecule.

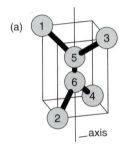

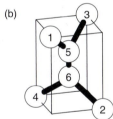

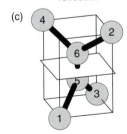

**Fig. 1.9** $S_4$ operations in $B_2Cl_4$.

The completed two-step operation which takes an atom at position 1, for example, to position 2 corresponds to $S_4^1$, and the associated operations $S_4^2$ and $S_4^3$ would take an atom at 1 to positions 3 and 4 respectively, with each stage involving the 90° rotation followed by reflection. However, when an $S_4$ axis is present, there is always a coincident $C_2$ axis, and the operation $S_4^2$ is regarded as $C_2$ in a similar way to $C_4^2$. Rotation–reflection axes $S_n$ with $n > 2$ are always found coincident with a normal rotation axis of the same, or lower, order.

The axis $S_1$ effectively involves just the reflection, and is therefore equivalent to a plane, whilst $S_2$ corresponds to the next symmetry element—the centre of symmetry.

## 1.6 Centre of symmetry: symbol i

The centre of symmetry in a structure—sometimes called the centre of inversion—is a specific point through which it is possible to project every other point in a structure to an equivalent position on the opposite side of the centre. In mathematical terms, a centre of symmetry will exist at the coordinate origin $(0, 0, 0)$ if, for every point $(x, y, z)$ in the structure, there exists an equivalent point at $(-x, -y, -z)$.

Molecules which contain a centre of symmetry are termed *centrosymmetric molecules*, and we shall see later that the presence of this symmetry element is an important distinguishing feature when considering the spectroscopic properties of such molecules.

This symmetry element is present in many highly symmetric structures, such as the square (Fig. 1.7) and regular hexagon (Fig. 1.8). It often occurs together with planes and axes, but it can also exist alone. Figure 1.10 shows a staggered, ethane-like structure in which different atoms X, Y and Z have been substituted for hydrogen. No axes or planes are to be found, but the centre remains, with atoms C1, X1, Y1 and Z1 related to C2, X2, Y2 and Z2 by inversion through the *centre* located at the middle of the C–C bond. A discrete shape can only have one centre of symmetry.

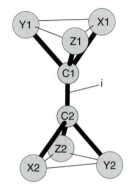

**Fig. 1.10**

## 1.7 The identity: symbol E

This final type of symmetry element is present in all shapes, and the associated operation corresponds to leaving the object in an identical configuration to the starting point. The simplest way to achieve this is to do nothing, but the identity is also equivalent to carrying out rotations such as $C_n^n$, as noted above, or the operation $C_1$. As we shall see in Chapter 2, its inclusion in this list of symmetry element arises from the requirements of group theory, and its importance is first encountered in this Primer when considering sequential symmetry operations and the construction of multiplication tables. Figure 1.11 shows the shape of a tetrahedral species MABCD in which the only element of symmetry present is the identity E.

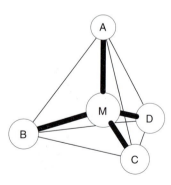

**Fig. 1.11**

## 1.8 Combinations of symmetry elements: the point group symbol

It will be clear from the examples above that although some molecules adopt shapes with very little symmetry, for others we can identify a variety of axes and planes, and perhaps also a centre of inversion. The two diagrams which make up Fig. 1.12 show all the symmetry elements present in a square-planar molecule such as $XeF_4$, and it is clear that if we wished to describe this structure without resorting to a model or a picture, it would be quite cumbersome. Ideally, what is required is a concise symbol, rather like an acronym, which would contain all the relevant symmetry information, and from which it would be possible to reconstruct the shape of $XeF_4$.

The problem is made considerably easier by the fact that for some shapes, not all the symmetry elements are independent. For example, in the diamond shape depicted in Fig. 1.4, it can be shown that the presence of any two of the mutually perpendicular $C_2$ axes automatically generates the third one. Similar relationships exist between other combinations of symmetry elements, and some of these will be illustrated later in Chapter 2. Their existence goes a considerable way towards simplifying what is known as the *point group symbol*.

This is the symbol which identifies and encapsulates the minimum number and type of symmetry elements needed to describe a particular shape.

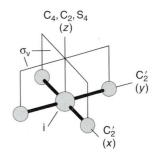
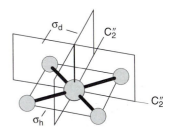

**Fig.1.12** Symmetry elements in $XeF_4$.

### Shapes with one C axis

In general, the point group symbol is constructed around the existence of the highest order proper rotation axis, and essential information about the additional presence of planes is included as a further subscript.

Figure 1.3 shows the shape of the $PCl_3$ molecule, with its $C_3$ axis and three vertical planes, but apart from the identity E, there are no other symmetry elements present. Figure 1.2 shows that it is possible to create a shape which possesses a $C_3$ axis, *without* three vertical planes, and so for $PCl_3$, the point group symbol must contain specific reference both to the $C_3$ axis and the vertical planes. The appropriate symbol is $C_{3v}$.

As indicated above, this symbol is derived from the presence of the single $C_3$ axis, to which has been added the subscript (v) indicating the vertical planes. Note that it is only necessary to specify 'v' once, as the action of $C_3$ will automatically result in the presence of the other two.

Figure 1.13 shows the molecule $SO_2F_2$. Here, there is one $C_2$ axis, and the two planes are both 'vertical'. The point group is $C_{2v}$.

We are also now in a position to identify the point groups of all the planar shapes shown in Figs. 1.2(a)–(c). These shapes were constructed to illustrate the proper rotation axes $C_n$ ($n = 2$–4), but because they are all planar, each shape has a $\sigma_h$ (the plane of the page) and a corresponding $S_n$ axis coincident with $C_n$.

The planar $C_n$ shapes with even $n$ have a centre of symmetry, but apart from this, there are no other symmetry elements present other than E and $\sigma_h$. $C_n$ takes precedence over $S_n$ in defining the point group symbol,

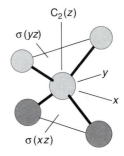

**Fig. 1.13**

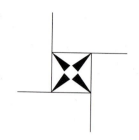

**Fig. 1.14**

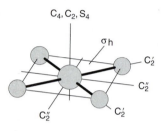

$C_4, C_2, S_4$

$\sigma_h$

$C_2'$

$C_2''$

$C_2''$

$C_2'$

**Fig.1.12′**
Symmetry axes in XeF$_4$.

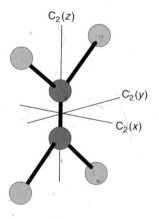

$C_2(z)$

$C_2(y)$

$C_2'(x)$

**Fig. 1.15**

and these shapes are therefore identified by the symbols $C_{nh}$, with $n = 2$–4. If the plane of symmetry in these shapes is destroyed—as in Fig. 1.14 (cf. Fig. 1.2)—the point group symbol reduces to just $C_n$.

**Shapes with more than one C axis: the symbol $D_n$**

Point group symbols of the type $C_{nv}$ or $C_{nh}$ apply to a large number of molecules, but are inappropriate when there is more than one symmetry axis present.

In a square-planar molecule such as XeF$_4$ (Fig. 1.12′), there is a single 'high order' $C_4$ axis (known as the principal axis), which defines 'vertical', and coincident with this are a $C_2$ and $S_4$. There are also four $C_2$ axes lying in the horizontal plane which are of two types: two lie along the Xe–F bonds ($C_2'$) whilst two lie between them ($C_2''$). The positions of these axes are shown alongside.

In order to make the description of multiple axes more concise, a new symbol is therefore introduced which recognises that '$n$' $C_2$ axes can happily coexist with one high order axis $C_n$ ($n > 2$), provided that they lie in a plane perpendicular to $C_n$. In XeF$_4$, the plane which contains these four $C_2$ axes is itself a plane of symmetry ($\sigma_h$), but, in general, it is not a requirement that it should be so, as will be seen later.

The new symbol which takes account of these axes is denoted by $D_n$, and it is defined as $D_n = C_n + n\perp C_2$ axes.

**Point group symbols $D_{nh}$ and $D_{nd}$**

The symbol $D_n$ successfully combines multiple axis information for all but a few very high symmetry shapes, and the presence of planes of symmetry is indicated by the further subscripts 'h' or 'd'.

Here, as previously, 'h' denotes horizontal, and indicates the presence of a plane of symmetry perpendicular to the principal axis. The subscript 'd' indicates the presence of a set of *dihedral* planes. These are vertical planes (i.e. they contain the principal axis), but are denoted by 'd' rather than 'v' to indicate that they bisect the $C_2$ axes perpendicular to the principal axis.

The molecule C$_2$H$_4$ (Fig. 1.15) and the simple rhombus (Fig. 1.4) are examples of shapes belonging to the point group $D_{2h}$. If we select one of the $C_2$ axes—$C_2(z)$—as the principal axis defining 'vertical', there is then a horizontal plane of symmetry $\sigma(xy)$. This combination of three $C_2$ axes and the horizontal plane is sufficient to establish the point group.

The square (Fig. 1.7) and XeF$_4$ (Fig. 1.12) have $D_{4h}$ symmetry, and the regular hexagon (Fig. 1.8) belongs to the point group $D_{6h}$.

The B$_2$Cl$_4$ molecule shown in Fig. 1.16 is an example of a shape with the point group symbol $D_{2d}$. In addition to the $C_2$ (and $S_4$) lying along the B–B bond, there are two $C_2$ axes perpendicular to this bond—denoted by $C_2'$ in the figure—and two dihedral planes which lie between the $C_2'$ axes.

## Shapes with several high order axes: special point groups

The $D_{nh}$ and $D_{nd}$ point groups illustrated above contain only one high order axis and $n \perp C_2$ axes, but for some highly symmetrical structures, it is possible for several high order ($n > 2$) axes to be present. Two of the most obvious shapes for which this is the case are the cube and the regular tetrahedron, to which may also be added the octahedron and the icosahedron.

These shapes possess a large number of symmetry elements, and it is no longer convenient to use a systematic combination of axes and planes to form the point group symbol. Instead, the point groups of these shapes are identified by specific symbols. The regular tetrahedron is assigned the point group symbol $T_d$, the cube and octahedron the symbol $O_h$, and the icosahedron the symbol $I_h$.

Figures 1.17 and 1.18 show the relationships between the tetrahedron, cube and octahedron, and identify the types of symmetry element present in the $T_d$ and $O_h$ point groups. Further discussion of these symmetry elements is found below:

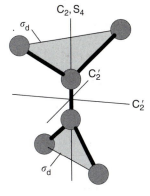

**Fig. 1.16**

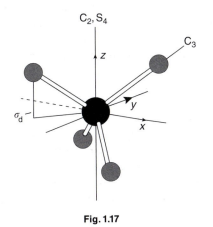

**Fig. 1.17**

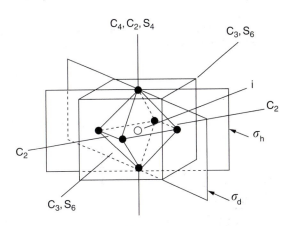

**Fig. 1.18**

## Point group symbols $C_{\infty v}$ and $D_{\infty h}$

The only important class of point groups which remain to be identified are those adopted by linear molecules, and by objects such as a wine glass (Fig. 1.19) or plate, which possess circular axial symmetry. For such shapes, it is realistic to describe the principal axis as being $C_\infty$ and to define this axis as 'vertical'. As all rotational positions about this axis are equivalent, an infinite number of vertical planes are also present.

For an object such as the wine glass, or for a linear molecule such as nitrous oxide (N–N–O), there are no other symmetry elements present apart from the identity, and the point group is therefore $C_{\infty v}$.

Some linear molecules possess, in addition, a centre of symmetry, a horizontal plane of symmetry (which contains the centre), and an infinite

**Fig. 1.19**

o–c–o

**Fig. 1.20**

number of $C_2$ axes lying in the horizontal plane, which also pass through the centre. The corresponding point group is $D_{\infty h}$, and Fig. 1.20 shows the shape of $CO_2$, which is a typical linear molecule with this symmetry.

## 1.9   Systematic point group classification

The point groups which have been encountered so far have been essentially illustrative, and rather selective. However, we are now in a position to set up a strategy for identifying the point group of any molecule, once we have identified the symmetry elements present. Such a scheme is given in Fig. 1.21.

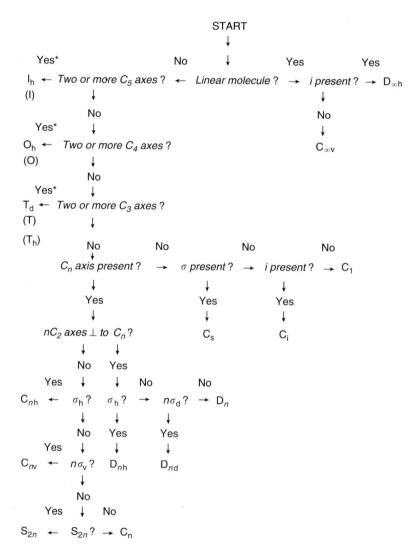

*Note: The pure rotation groups I, O and T would also be extracted at these points, but very few, if any, molecules are known with these symmetries. Molecules belonging to the $T_h$ point group are very rare. They may be distinguished from those with $T_d$ symmetry by the presence of a centre, i.

**Fig. 1.21** Scheme for identifying the point group of any molecule.

The examples which appear at the end of this chapter offer several opportunities to become familiar with this process, but it is useful to consider briefly here the operation of this flow diagram for the eclipsed form of $C_2H_6$, reproduced in Fig. 1.22(a).

Firstly, the molecule is not linear, and inspection shows that it only contains one $C_3$ axis. The next stage is to check for perpendicular $C_2$ axes. These are clearly present, and the point group symbol therefore starts $D_3$ ... This eclipsed structure also possesses a $\sigma_h$, and the point group is therefore $D_{3h}$. This structure also possesses three planes $\sigma_v$, one of which is shown here, but it is not necessary to include it in the route to the point group.

The staggered form (Fig. 1.22(b)) also exhibits $C_3$ and $C_2$ axes, but *has no horizontal plane*. The existence of three dihedral planes $\sigma_d$ (one is shown here, bisecting two of the $C_2$ axes) identifies the point group as $D_{3d}$. These examples illustrate a general relationship between staggered and eclipsed structures which involve one principal $n$-fold axis. The eclipsed form of the structure will belong to the point group $D_{nh}$, whilst the staggered form will belong to $D_{nd}$. In addition, it may be shown that a centre of symmetry is present in all $D_{nh}$ point groups when $n$ is even, and in all $D_{nd}$ point groups when $n$ is odd.

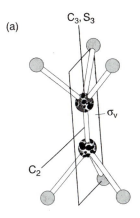

(a) $C_3, S_3$

$\sigma_v$

$C_2$

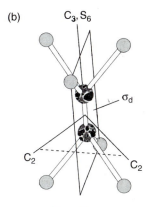

(b) $C_3, S_6$

$\sigma_d$

$C_2$    $C_2$

**Fig. 1.22**

## 1.10 A first look at character tables

Examination of the flow diagram in Fig. 1.21 shows that the majority of decisions to be made rest on the location of planes and proper axes of rotation, and that only very rarely it is necessary to identify all the symmetry elements present in order to arrive at the correct point group. This is an obvious advantage, but one may then overlook certain symmetry elements—notably any $S_n$ axes—when needing to compile a complete list.

It is therefore useful at this stage to take a first look at *character tables*. Appendix II includes most of the character tables necessary for normal chemical purposes, and the character table for the point group $C_{2v}$ is reproduced below:

| $C_{2v}$ | $E$ | $C_2$ | $\sigma_v(xz)$ | $\sigma_v'(yz)$ | $h = 4$ | |
|---|---|---|---|---|---|---|
| $A_1$ | 1 | 1 | 1 | 1 | $z$ | $x^2, y^2, z^2$ |
| $A_2$ | 1 | 1 | $-1$ | $-1$ | $R_z$ | $xy$ |
| $B_1$ | 1 | $-1$ | 1 | $-1$ | $x, R_y$ | $xz$ |
| $B_2$ | 1 | $-1$ | $-1$ | 1 | $y, R_x$ | $yz$ |

The body of a typical character table generally appears as rows and columns of numbers, which we shall come to use later in this Primer. However, the very top line of symbols contains information relating to the symmetry elements and operations present in the point group, and at this stage this row of symbols can be used as a prompt to check that all the elements present can be identified.

This row also focuses attention on the distinction between a symmetry element and its operation(s), and at the end of this top row, some character tables include a value for '$h$', which is known as the *order* of the group. This value of '$h$' corresponds to the *sum of the coefficients* of the symmetry elements listed in the top row, and it also happens to be the maximum number of points generated by the symmetry operations of the group acting on a single initial point.

Reproduced below are the top rows of six frequently encountered character tables:

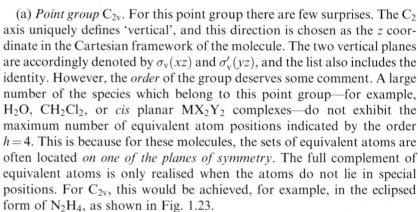

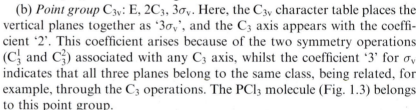

(a)  $C_{2v}$  E   $C_2$       $\sigma_v(xz)$  $\sigma_v'(yz)$   $h=4$
(b)  $C_{3v}$  E   $2C_3$      $3\sigma_v$   $h=6$
(c)  $C_{4v}$  E   $2C_4$      $C_2$   $2\sigma_v$   $2\sigma_d$       $h=8$
(d)  $D_{3h}$  E   $2C_3$      $3C_2$   $\sigma_h$    $2S_3$        $3\sigma_v$   $h=12$
(e)  $T_d$     E   $8C_3$      $3C_2$   $6S_4$   $6\sigma_d$       $h=24$
(f)  $O_h$     E   $8C_3$      $6C_2$   $6C_4$   $3C_2(=C_4^2)$   i   $6S_4$   $8S_6$   $3\sigma_h$   $6\sigma_d$   $h=48$

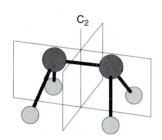

$C_2$

**Fig. 1.23**

(a) *Point group* $C_{2v}$. For this point group there are few surprises. The $C_2$ axis uniquely defines 'vertical', and this direction is chosen as the $z$ coordinate in the Cartesian framework of the molecule. The two vertical planes are accordingly denoted by $\sigma_v(xz)$ and $\sigma_v'(yz)$, and the list also includes the identity. However, the *order* of the group deserves some comment. A large number of the species which belong to this point group—for example, $H_2O$, $CH_2Cl_2$, or *cis* planar $MX_2Y_2$ complexes—do not exhibit the maximum number of equivalent atom positions indicated by the order $h=4$. This is because for these molecules, the sets of equivalent atoms are often located *on one of the planes of symmetry*. The full complement of equivalent atoms is only realised when the atoms do not lie in special positions. For $C_{2v}$, this would be achieved, for example, in the eclipsed form of $N_2H_4$, as shown in Fig. 1.23.

(b) *Point group* $C_{3v}$: E, $2C_3$, $3\sigma_v$. Here, the $C_{3v}$ character table places the vertical planes together as '$3\sigma_v$', and the $C_3$ axis appears with the coefficient '2'. This coefficient arises because of the two symmetry operations ($C_3^1$ and $C_3^2$) associated with any $C_3$ axis, whilst the coefficient '3' for $\sigma_v$ indicates that all three planes belong to the same class, being related, for example, through the $C_3$ operations. The $PCl_3$ molecule (Fig. 1.3) belongs to this point group.

(c) *Point group* $C_{4v}$: E, $2C_4$, $C_2$, $2\sigma_v$, $2\sigma_d$. Figure 1.24 shows a typical structure with $C_{4v}$ symmetry, as adopted by a square pyramidal molecule such as $XeOF_4$. Coincident with the $C_4$ principal axis is the $C_2$ axis whose operation is equivalent to $C_4^2$, and there are two sets of vertical planes. One pair ($\sigma_v$) contains the basal fluorine atoms whilst the second pair—denoted by '$\sigma_d$' lies between them.

The character table identifies both the $C_4$ and coincident $C_2$ as independent symmetry axes, and reinforces this by taking account of only two of the $C_4$ operations, $C_4^1$ and $C_4^3$, through the term '$2C_4$'. However, the nomenclature given to the two sets of planes is not immediately justifiable.

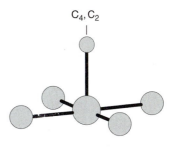

Both the $\sigma_v$ and $\sigma_d$ planes are clearly vertical, but this point group contains no perpendicular $C_2$ axes which we have come to associate with the term 'dihedral'. At first glance, these $\sigma_d$ planes are just a second set of vertical planes, and could accordingly be labelled $\sigma_v'$.

However, although this point group does not contain any $C_2$ *axes* perpendicular to the principal axis, molecules with this symmetry have their *Cartesian axes* $x$ and $y$ perpendicular to $C_4$, and these axes may be used to distinguish the two sets of vertical planes. The planes denoted by $\sigma_v$ are identified with the Cartesian planes $xz$ and $yz$, whilst the $\sigma_d$ planes bisect the $x$ and $y$ axes, as shown.

(d) *Point group* $D_{3h}$: E, $2C_3$, $3C_2$, $\sigma_h$, $2S_3$, $3\sigma_v$. The defining features of this point group—a $C_3$ axis with three $C_2$ axes perpendicular to it, and a horizontal symmetry plane—require little further discussion, except to note that once again account is taken of the two operations $C_3^1$ and $C_3^2$. The three vertical planes belong to a common class, and are grouped together. The symmetry element which is most easily missed here is the $S_3$ axis, which is coincident with the $C_3$ axis. As we saw earlier, the eclipsed form of $C_2H_6$ (Fig. 1.22(a)) would belong to this point group. Figure 1.25 shows a further example of a structure with this symmetry—the trigonal bipyramid.

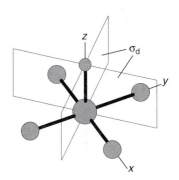

(e) *Point group* $T_d$: E, $8C_3$, $3C_2$, $6S_4$, $6\sigma_d$. This point group, although containing fewer symmetry elements than a genuine cube, is regarded as a 'cubic' point group because it contains four $C_3$ axes. These ensure, among other things, that molecules with this symmetry are isotropic.

The regular tetrahedral structure adopted by a molecule such as $CH_4$ has been shown in Fig. 1.17. The $T_d$ character table identifies the symmetry elements $C_3$, $C_2$, $S_4$ and $\sigma_d$ in addition to E, and the location of the axes and planes is indicated in this figure. The entry '$8C_3$' is made up of two factors. Firstly, there are four distinct $C_3$ axes each coincident with a C–H bond. These four axes belong to the same class, but each one has two associated operations, $C_3^1$ and $C_3^2$, giving a total of eight operations in this class.

There are three distinct $C_2$ axes lying along the Cartesian axes, and coincident with each of these is an $S_4$ axis. Each $S_4$ axis can generate two distinct operations, $S_4^1$ and $S_4^3$, and this accounts for the entry '$6S_4$' in the character table. Although there is no unique choice for the 'vertical' direction in such a symmetrical shape, it is convenient to set up the Cartesian coordinates along the directions of the $C_2$ axes. This has the result that the planes of symmetry—listed as '$6\sigma_d$'—each contain one of the $C_2$ axes but lie between the other two, so justifying the subscript 'd'.

It is important to note that the regular tetrahedron does not have a centre of symmetry.

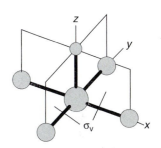

**Fig. 1.24**

(f) *Point group* $O_h$: E, $8C_3$, $6C_2$, $6C_4$, $3C_2$, i, $6S_4$, $8S_6$, $3\sigma_h$, $6\sigma_d$. The relationship between the regular octahedron and the cube has been encountered previously (see Fig. 1.18). Both these shapes belong to the $O_h$ point group, and octahedral molecules containing a central atom are to be found throughout the periodic table.

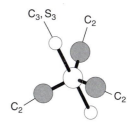

**Fig. 1.25**

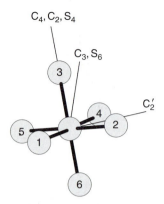

$C_4, C_2, S_4$

$C_3, S_6$

$C_2'$

**Fig. 1.26**

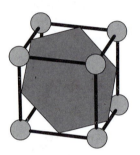

**Fig. 1.27**

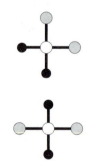

**Fig.1.28**
*cis* and *trans* isomers of $MX_2Y_2$

Figure 1.26 shows the location of each axis type in an octahedral molecule such as $SF_6$. The entries '$8C_3$', '$6C_4$' and '$6S_4$' take account of the fact that each $C_3$, $C_4$ and $S_4$ axis is associated with two distinct operations. The entry '$3C_2$' refers to the $C_2$ axes which are coincident with the $C_4$ axes. The entry '$6C_2$' indicates a second type of $C_2$ axis passing through opposite *edges* of the cube or octahedron. In Fig. 1.26, such an axis is denoted by $C_2'$.

The remaining entry '$8S_6$' is the most difficult to account for. Coincident with each of the $C_3$ axes is an $S_6$ rotation–reflection axis, which has associated with it the six operations $S_6^n$, with $n = 1$–6. Of these, $S_6^2$ and $S_6^4$ correspond to $C_3^1$ and $C_3^2$. $S_6^3$ is the same as the operation i and $S_6^6 = E$. This leaves only $S_6^1$ and $S_6^5$ as distinct operations not accounted for elsewhere, and the existence of these two operations for each of the four $S_6$ axes accounts for the entry '$8S_6$'. In Fig. 1.26, the operations performed using the $S_6$ axis indicated would be to shift an atom at position 1 sequentially to positions 6, 2, 4, 3 and 5 before returning to position 1. This six-fold rotational symmetry also provides a clue as to how one might bisect a solid cube or octahedron to leave each piece showing a hexagonal face. The solution to this puzzle for the cube is indicated in Fig. 1.27.

*Note on terminology.* The high symmetry $T_d$ and $O_h$ point groups discussed above place considerable constraints on molecular structure. However, the terms 'tetrahedral' and 'octahedral' are also used more generally throughout chemistry, and it is important to retain the distinction, for example, between strict $T_d$ symmetry and a more relaxed usage, which would describe a molecule with $C_{3v}$ symmetry, such as $CH_3Cl$, as also being tetrahedral. $CH_3Cl$ is indeed tetrahedral, but it is not a *regular* tetrahedron. A related problem can arise over the description 'octahedral'. A large number of transition metal complexes are octahedral, or contain an octahedrally coordinated metal centre within a much larger structure. In the majority of cases, this does not imply $O_h$ symmetry, but rather that there are six ligands distributed approximately uniformly within the first coordination shell.

The term 'square planar' is similarly used almost universally to describe planar complexes containing four ligands, without the limitations imposed by strict $D_{4h}$ symmetry. For example, the *cis* and *trans* isomers of molecules such as $PtX_2Y_2$, where X and Y may be halogen or nitrogen atom donors, are likely to belong to the point groups $C_{2v}$ and $D_{2h}$ respectively, as may be deduced from Fig. 1.28.

This wider usage need not cause problems, but should be borne in mind when circumstances require a more precise symmetry description.

### 1.11 Point groups of chiral molecules

Molecules which can induce rotation of plane polarised light are termed chiral, and can exist in at least two optically active isomeric forms, called enantiomers. The basic requirement for optical activity is that the molecule must have no planes of symmetry, and no centre of symmetry. A classic shape which exhibits this phenomenon would be a methane derivative such as CHFBrCl. Here, there are no symmetry elements apart from E, and the molecule belongs to the point group $C_1$. The more general tetrahedral shape MABCD has appeared previously as Fig. 1.11.

Although optical isomers are most frequently associated with this complete asymmetry, there are in fact several other point groups where chirality may be found.

Figure 1.29 shows a stylised three-bladed propeller together with its mirror image. These two shapes are not superposable, and are therefore optical isomers, existing in obvious left- and right-handed forms. This shape, however, possesses a surprising amount of symmetry, containing not only a principal $C_3$ axis but also three perpendicular $C_2$ axes, leading to the point group symbol $D_3$.

The molecular analogue of this shape would be a tris-chelate molecule based on octahedral coordination, and the essential framework for such a structure is shown in Fig. 1.30. Typical chelate groups which can form strong bridges between *cis* positions in the octahedron are ethylene diamine (en) or the oxalate group (ox), with the result that optically active complexes containing $[Co(en)_3]^{3+}$ or $[Fe(ox)_3]^{3-}$ are readily formed.

This point group is also related to the eclipsed and staggered configurations described previously for ethane. The eclipsed form has symmetry $D_{3h}$, and if one $CH_3$ group is rotated relative to the other by 60°, the structure becomes staggered, with point group $D_{3d}$. Relative rotation by any angle between 0° and 60° destroys the vertical symmetry planes, but the three perpendicular $C_2$ axes are retained, leading to the same $D_3$ point group.

In general, all the $D_n$ point groups are expected to show chirality, and the same would be true for lower symmetry point groups such as $C_n$. In practice, however, unless there is a rigid framework which prevents free rotation, it is often impossible to isolate the separate enantiomers.

Thus, for a molecule such as $H_2O_2$, which has $C_2$ symmetry, facile rotation about the O–O bond would effectively convert one isomer into the other, preventing isolation of a unique configuration. However, in a more rigid $C_2$ structure involving chelate groups, such as is shown in Fig. 1.31, it should be possible to resolve the optical isomers.

## 1.12 Summary

The principal aim of this chapter has been to illustrate the various symmetry elements and operations which can lead to a description of molecular symmetry in terms of the point group. In addition, it is hoped that at least some of the examples given will encourage a wider perception and curiosity in this subject.

The problems and exercises which follow are designed to provide additional practice in identifying symmetry elements and point groups. The majority of the problems have only one correct answer, but a few of

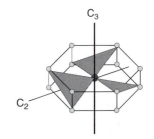

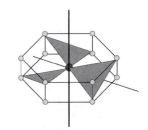

**Fig. 1.29**

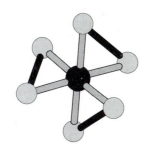

**Fig. 1.30**

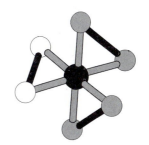

**Fig. 1.31**

the exercises are more open-ended, and will hopefully stimulate discussion in a tutorial environment.

### 1.13  Problems and exercises

1.  Identify the point groups of the following molecules:
(a) $BF_3$, (b) $SF_5Cl$, (c) $CH_3Br$, (d) *trans* $CHCl=CHCl$,

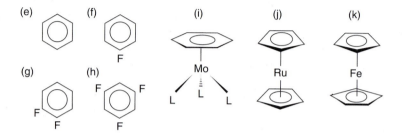

2.  Identify which of the above structures contain (i) a centre of inversion, (ii) a rotation–reflection axis. In the latter case, state the order of axis present.
3.  Sketch an object with $D_4$ symmetry and also its optical isomer.
4.  Sketch an object which would have $C_{9h}$ symmetry. How would you modify your shape such that the symmetry is reduced to $C_9$?

# 2 Matrices, multiplication tables and representations

The pictorial illustrations of symmetry operations shown in Chapter 1 describe single operations quite satisfactorily. However, they can become cumbersome if we need to consider sequential operations and, more importantly, they provide no stepping-off point if we should wish to relate molecular symmetry to topics such as molecular energy levels and spectroscopy.

For this, we require a description of symmetry operations in more mathematical language, and the first step is to be able to describe and unify symmetry operations through the use of matrices.

## 2.1 Introduction to matrices—some definitions

Matrices are arrays of either numbers or variables, and come in various shapes and sizes. The four basic shapes are the *column matrix*, the *row matrix*, the *square matrix* and the *rectangular matrix*, and examples of each of these are given alongside. Each number or variable in the matrix is called a *matrix element*, and its position is described by a grid system in which rows are numbered first, starting at the top, and columns second, starting at the left-hand side.

$$\mathbf{W} = \begin{pmatrix} W_{11} \\ W_{21} \\ W_{31} \end{pmatrix}$$

$$\mathbf{X} = (X_{11} \ X_{12} \ X_{13})$$

Thus matrix $\mathbf{W}$ is a column matrix, with three elements $W_{11}$, $W_{21}$ and $W_{31}$, and matrix $\mathbf{X}$ is a row matrix, with elements $X_{11}$, $X_{12}$ and $X_{13}$. The square matrix shown as $\mathbf{Y}$ has a total of nine elements, and following our grid system, we can identify the elements $Y_{22}$ and $Y_{31}$, for example, as the integers 4 and 6 respectively. In the same way, $\mathbf{Z}$ is a rectangular matrix, having a total of six elements, as shown.

$$\mathbf{Y} = \begin{pmatrix} 1 & 2 & 5 \\ 3 & 4 & 7 \\ 6 & 9 & 8 \end{pmatrix}$$

$$\mathbf{Z} = \begin{pmatrix} a & b \\ c & d \\ e & f \end{pmatrix}$$

Matrices have a wide and general application in many areas of science, and Appendix III lists a number of additional sources of material which describe the properties of matrices in detail.

The only types of matrix encountered in this Primer are column matrices and square matrices, and we shall use them to describe symmetry operations. In order to do this, we shall need to know basically only two further definitions—the *diagonal* of a matrix and the *character* of a matrix—together with the rules which govern matrix multiplication.

We shall encounter the *diagonal* and the *character* only in square matrices, and although a square can in general have two diagonals, the term *diagonal* when applied specifically to a matrix refers to the leading diagonal, which for a matrix $\mathbf{M}$ is defined by elements such as $M_{11}$, $M_{22}$, $M_{33}$, etc.—i.e. top left to bottom right.

$$\mathbf{M} = \begin{pmatrix} M_{11} & & \\ & M_{22} & \\ & & M_{33} \end{pmatrix}$$

The *character* of a matrix is the sum of the elements on the diagonal, and is given the symbol $\chi$. In matrix **Y** above, $\chi$ has the value 13.

## 2.2    Matrix multiplication

Matrix multiplication is, at first sight, a complicated procedure when compared with the simple multiplication of numbers. Firstly, not all matrices can be multiplied together, and secondly, for those that can, the result of such an operation—typically a third matrix—must involve all the matrix elements in the two initial matrices.

The rule which defines matrix multiplication can be stated very generally for two matrices **A** and **B** whose product is a third matrix **C**. If we identify each element in matrix **C** as $C_{ij}$, where $i$ and $j$ refer to the row and column of the element respectively, then the value of $C_{ij}$ is obtained as

$$C_{ij} = \sum A_{ik} \cdot B_{kj} \quad \text{for the process} \quad \mathbf{A} \times \mathbf{B} = \mathbf{C}$$

In this expression, the summation is carried out from $k = 1$ to the maximum value of $k$, as defined by the size of the matrices.

As an example, let us consider the multiplication of the two $2 \times 2$ square matrices **P** and **Q** to give a new matrix **R**:

$$\mathbf{P} = \begin{pmatrix} 1 & 2 \\ 3 & 1 \end{pmatrix}$$

$$\mathbf{Q} = \begin{pmatrix} 3 & 1 \\ -2 & 0 \end{pmatrix}$$

$$\mathbf{R} = \begin{pmatrix} -1 & 1 \\ 7 & 3 \end{pmatrix}$$

$$R_{11} = P_{11} \cdot Q_{11} + P_{12} \cdot Q_{21} = 1 \times 3 + 2 \times (-2) = -1$$

$$R_{12} = P_{11} \cdot Q_{12} + P_{12} \cdot Q_{22} = 1 \times 1 + 2 \times 0 = 1$$

$$R_{21} = P_{21} \cdot Q_{11} + P_{22} \cdot Q_{21} = 3 \times 3 + 1 \times (-2) = 7$$

$$R_{22} = P_{21} \cdot Q_{12} + P_{22} \cdot Q_{22} = 3 \times 1 + 1 \times 0 = 3$$

Note that this operation has assumed a *particular order* for carrying out this multiplication, i.e. $\mathbf{P} \times \mathbf{Q}$. The alternative order $\mathbf{Q} \times \mathbf{P}$ would result in matrix elements such as

$$R'_{11} = Q_{11} \cdot P_{11} + Q_{12} \cdot P_{21} = 3 \times 1 + 1 \times 3 = 6$$

which, for this example, is clearly different from $R_{11}$.

This illustrates an important difference between the multiplication of *matrices* and the multiplication of *numbers*. Simple multiplication is *commutative*, i.e. $2 \times 3 = 3 \times 2$, but matrix multiplication is only commutative if both the matrices are symmetrical about the leading diagonal.

## 2.3    Symmetry operations and matrices

We are now in a position both to express symmetry operations in matrix notation, and also to examine the consequences of sequential symmetry operations—again through the use of matrices.

## Matrices for $C_2$ rotations

As an initial example, we may consider the effect of two-fold rotations around the Cartesian axes on a general point $P(X, Y, Z)$ as shown in Fig. 2.1.

If the $C_2(x)$ axis is coincident with $x$, then the effect of $C_2(x)$ on P is to move this point to the position $Q$, with new coordinates $(X, -Y, -Z)$.

The effect of this rotation can be expressed as three separate equations

$$C_2(x)X = (+1)X + (0)Y + (0)Z$$
$$C_2(x)Y = (0)X + (-1)Y + (0)Z$$
$$C_2(x)Z = (0)X + (0)Y + (-1)Z$$

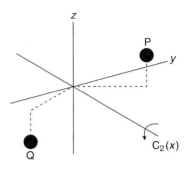

**Fig. 2.1**

and also in matrix form, where the correspondence with the separate equations above is clearly apparent:

$$C_2(x)\begin{pmatrix} X \\ Y \\ Z \end{pmatrix} = \begin{pmatrix} 1 & 0 & 0 \\ 0 & -1 & 0 \\ 0 & 0 & -1 \end{pmatrix}\begin{pmatrix} X \\ Y \\ Z \end{pmatrix} = \begin{pmatrix} X \\ -Y \\ -Z \end{pmatrix}$$

In a similar way, $C_2$ rotation about the $y$ axis (Fig. 2.2) shifts $(X, Y, Z)$ to the new position $R$ at $(-X, Y, -Z)$ and is expressed as

$$C_2(y)\begin{pmatrix} X \\ Y \\ Z \end{pmatrix} = \begin{pmatrix} -1 & 0 & 0 \\ 0 & 1 & 0 \\ 0 & 0 & -1 \end{pmatrix}\begin{pmatrix} X \\ Y \\ Z \end{pmatrix} = \begin{pmatrix} -X \\ Y \\ -Z \end{pmatrix}$$

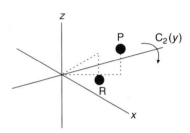

**Fig. 2.2**

whilst $C_2(z)$ acting on $P$ shifts the point to $S(-X, -Y, Z)$, as in Fig. 2.3:

$$C_2(z)\begin{pmatrix} X \\ Y \\ Z \end{pmatrix} = \begin{pmatrix} -1 & 0 & 0 \\ 0 & -1 & 0 \\ 0 & 0 & 1 \end{pmatrix}\begin{pmatrix} X \\ Y \\ Z \end{pmatrix} = \begin{pmatrix} -X \\ -Y \\ Z \end{pmatrix}$$

Each of these matrices is therefore associated with a particular rotational symmetry operation. Matrices which represent the operations of higher order $C_n$ axes and $S_n$ axes will be encountered in Chapter 4.

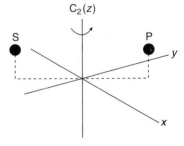

**Fig. 2.3**

## Matrices for reflections

As may be seen from Fig. 2.4, the operation of reflection in the plane $\sigma(xy)$ would take point $P(X, Y, Z)$ to position $T$ at $(X, Y, -Z)$ and, in matrix notation, this would be summarised as

$$\sigma(xy)\begin{pmatrix} X \\ Y \\ Z \end{pmatrix} = \begin{pmatrix} 1 & 0 & 0 \\ 0 & 1 & 0 \\ 0 & 0 & -1 \end{pmatrix}\begin{pmatrix} X \\ Y \\ Z \end{pmatrix} = \begin{pmatrix} X \\ Y \\ -Z \end{pmatrix}$$

The matrices for reflection in planes $\sigma(xz)$ and $\sigma(yz)$ are

$$\sigma(xz) = \begin{pmatrix} 1 & 0 & 0 \\ 0 & -1 & 0 \\ 0 & 0 & 1 \end{pmatrix}, \qquad \sigma(yz) = \begin{pmatrix} -1 & 0 & 0 \\ 0 & 1 & 0 \\ 0 & 0 & 1 \end{pmatrix}$$

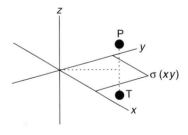

**Fig. 2.4**

$$E = \begin{pmatrix} 1 & 0 & 0 \\ 0 & 1 & 0 \\ 0 & 0 & 1 \end{pmatrix}$$

$$i = \begin{pmatrix} -1 & 0 & 0 \\ 0 & -1 & 0 \\ 0 & 0 & -1 \end{pmatrix}$$

### Matrices for E and i

The matrices which describe the identity and inversion operations on a general point $P(X, Y, Z)$ are very straightforward. The effect of the identity is to leave the coordinates unchanged, and operation of inversion through a centre located at the coordinate origin $(0, 0, 0)$ takes the point $(X, Y, Z)$ to the position $(-X, -Y, -Z)$.

The matrices for the operations E and i are shown alongside.

### Sequential symmetry operations in the point group $D_2$

We are now in a position to explore the effect of carrying out sequential symmetry operations both pictorially and by using simple matrices, and one of the easiest systems to visualise arises when all the important symmetry operations involve simple rotations.

Figure 2.5 shows the general point $P(X, Y, Z)$ in a Cartesian framework where a $C_2$ axis lies along each of the directions $x$, $y$ and $z$. This collection of symmetry elements, together with the identity, constitute the $D_2$ point group.

Rotation about $x$ via $C_2(x)$ takes P to a position Q, with coordinates $(X, -Y, -Z)$, and *subsequent rotation* about $y$ using $C_2(y)$ takes Q to the position S, at $(-X, -Y, Z)$.

If we then carry out the $C_2(z)$ operation on S, we return to the original position P.

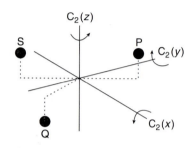

**Fig. 2.5**

The separate operations could be written as

$$C_2(x)P \rightarrow Q, \quad C_2(y)Q \rightarrow S, \quad C_2(z)S \rightarrow P$$

and the complete sequence of operations as

$$C_2(z) \cdot C_2(y) \cdot C_2(x)P \rightarrow P$$

Note that the first operation to be carried out, $C_2(x)$, appears at the *right-hand side* of the above sequence of symmetry operations.

These sequential operations illustrate one of the most important features of a point group: that the result of carrying out two or more operations in the group can in general be achieved separately by another operation of the group.

In this case, the result of the first two operations,

$$C_2(y) \cdot C_2(x)P \rightarrow S, \text{ can be achieved by } C_2(z)P \rightarrow S$$
$$\text{i.e. } C_2(y) \cdot C_2(x) = C_2(z)$$

The complete sequence $C_2(z) \cdot C_2(y) \cdot C_2(x)$ is equivalent to E.

### Sequential operations—use of matrices

The results above have been obtained by an essentially pictorial route, but the same conclusions can also be reached by multiplying together the *matrices* for the separate operations—as derived above.

For example, the sequential operations $C_2(y) \cdot C_2(x)$ would be obtained from the product of the appropriate matrices:

$$\begin{pmatrix} -1 & 0 & 0 \\ 0 & 1 & 0 \\ 0 & 0 & -1 \end{pmatrix} \begin{pmatrix} 1 & 0 & 0 \\ 0 & -1 & 0 \\ 0 & 0 & -1 \end{pmatrix} = \begin{pmatrix} -1 & 0 & 0 \\ 0 & -1 & 0 \\ 0 & 0 & 1 \end{pmatrix}$$

and the resulting matrix is seen to correspond to that for the operation $C_2(z)$.

Note that the matrices for $C_2(x)$ and $C_2(y)$ are both symmetrical about the leading diagonal, and because of this, these two operations will commute, i.e. $C_2(y) \cdot C_2(x) = C_2(x) \cdot C_2(y)$.

Successive rotations about the *same* $C_2$ axis, such as $C_2(y) \cdot C_2(y)$, result in the identity E. In matrix form, this arises from

$$\begin{pmatrix} -1 & 0 & 0 \\ 0 & 1 & 0 \\ 0 & 0 & -1 \end{pmatrix} \begin{pmatrix} -1 & 0 & 0 \\ 0 & 1 & 0 \\ 0 & 0 & -1 \end{pmatrix} = \begin{pmatrix} 1 & 0 & 0 \\ 0 & 1 & 0 \\ 0 & 0 & 1 \end{pmatrix}$$

### Multiplication table for the operations in point group $D_2$

In general, the effect of carrying out sequential symmetry operations can be expressed in the form of a *multiplication table*. The character table for the point group $D_2$ (Appendix II) has four symmetry elements, each of which is capable of only one operation:

$$D_2 \quad E \quad C_2(z) \quad C_2(y) \quad C_2(x)$$

The multiplication table for the $D_2$ point group is shown below:

| $D_2$ | E | $C_2(z)$ | $C_2(y)$ | $C_2(x)$ |
|---|---|---|---|---|
| E | E | $C_2(z)$ | $C_2(y)$ | $C_2(x)$ |
| $C_2(z)$ | $C_2(z)$ | E | $C_2(x)$ | $C_2(y)$ |
| $C_2(y)$ | $C_2(y)$ | $C_2(x)$ | E | $C_2(z)$ |
| $C_2(x)$ | $C_2(x)$ | $C_2(y)$ | $C_2(z)$ | E |

In this table, the *first operation* to be carried out is given along the *top row*, and the *second operation* is given in the left-hand *vertical column*. The body of the table then indicates the single symmetry operation which is equivalent to performing the two sequential operations.

This table reinforces the importance of the identity E in a symmetry group, and also shows that for this point group, all the operations are commutative: e.g. $C_2(x) \cdot C_2(y) = C_2(y) \cdot C_2(x) = C_2(z)$.

### Sequential operations in the $C_{2v}$ point group

Figure 2.6 shows the symmetry elements in the $C_{2v}$ point group in relation to the general point $P(X, Y, Z)$, and the multiplication table for $C_{2v}$ can be constructed either pictorially, or directly using matrices.

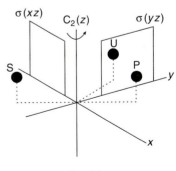

**Fig. 2.6**

As may be seen from this figure, a reflection in the $yz$ plane followed by a reflection in the $xz$ plane produces the sequence $P \rightarrow U \rightarrow S$, and single stage $P \rightarrow S$ could be achieved by $C_2$ rotation about $z$.

The matrices for these operations have already been obtained as

$$\begin{pmatrix} 1 & 0 & 0 \\ 0 & -1 & 0 \\ 0 & 0 & 1 \end{pmatrix} \quad \begin{pmatrix} -1 & 0 & 0 \\ 0 & 1 & 0 \\ 0 & 0 & 1 \end{pmatrix} \quad \begin{pmatrix} -1 & 0 & 0 \\ 0 & -1 & 0 \\ 0 & 0 & 1 \end{pmatrix}$$

$$\sigma(xz) \qquad\qquad \sigma(yz) \qquad\qquad C_2(z)$$

and, in matrix form, the result of a sequential operation such as $\sigma(xz) \cdot \sigma(yz)$ is therefore

$$\begin{pmatrix} 1 & 0 & 0 \\ 0 & -1 & 0 \\ 0 & 0 & 1 \end{pmatrix} \begin{pmatrix} -1 & 0 & 0 \\ 0 & 1 & 0 \\ 0 & 0 & 1 \end{pmatrix} = \begin{pmatrix} -1 & 0 & 0 \\ 0 & -1 & 0 \\ 0 & 0 & 1 \end{pmatrix}$$

i.e. $\sigma(xz) \cdot \sigma(yz) = C_2(z)$.

The complete multiplication table for $C_{2v}$ is shown below:

| $C_{2v}$ | E | $C_2(z)$ | $\sigma_v(xz)$ | $\sigma_v'(yz)$ |
|---|---|---|---|---|
| E | E | $C_2(z)$ | $\sigma_v(xz)$ | $\sigma_v'(yz)$ |
| $C_2(z)$ | $C_2(z)$ | E | $\sigma_v'(yz)$ | $\sigma_v(xz)$ |
| $\sigma_v(xz)$ | $\sigma_v(xz)$ | $\sigma_v'(yz)$ | E | $C_2(z)$ |
| $\sigma_v'(yz)$ | $\sigma_v'(yz)$ | $\sigma_v(xz)$ | $C_2(z)$ | E |

## 2.4   The concept of a representation

In the multiplication tables for $D_2$ and $C_{2v}$ above, the individual operations appear in symbolic form, i.e. E, $C_2(z)$ etc. but the multiplication table is also true if instead of using these symbols, we replace them by the *matrices* which we have used above to *represent* the various symmetry operations.

The fact that we can use a group of matrices in this way to represent symmetry operations leads to the general concept of a *representation* as being a set of matrices which can be placed in correspondence with the symmetry operations of the group, and which obey the group multiplication table.

However, the matrices which determine the fate of $P(X, Y, Z)$ in the examples above are *not unique* in having this property, and nor are they the simplest sets of matrices which behave in this way.

Figure 2.7 shows the elements of symmetry in the $C_{2v}$ point group, and can be used to study the effect of the operations on the single vector $x_1$ as shown. The identity E and the operation $\sigma(xz)$ leave this vector unaffected, but $\sigma(yz)$ and $C_2(z)$ reverse its direction, and these results

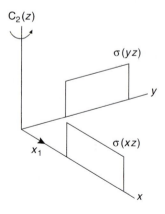

**Fig. 2.7**

can be summarised as

$$E(x_1) \rightarrow (x_1), \quad C_2(z)(x_1) \rightarrow -(x_1),$$

$$\sigma(xz)(x_1) \rightarrow (x_1), \quad \sigma(yz)(x_1) \rightarrow -(x_1)$$

Here, the *coefficients* of $x_1$ on the right-hand side of each of these expressions can be thought of as the *$1 \times 1$ matrices* $(+1)$, $(-1)$, $(+1)$ and $(-1)$ which represent the respective symmetry operations, in a similar way to the $3 \times 3$ matrices which describe the movement of point $P(X, Y, Z)$.

These numbers also obey the $C_{2v}$ multiplication table, as can be seen when the symmetry operations in $C_{2v}$ are replaced by the corresponding coefficients: i.e.

$$E = +1, \quad C_2(z) = -1, \quad \sigma_v(xz) = +1, \quad \sigma_v'(yz) = -1$$

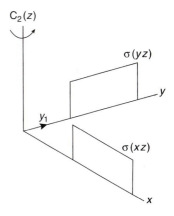

| $C_{2v}$ | E | $C_2(z)$ | $\sigma_v(xz)$ | $\sigma_v'(yz)$ |
|----------|-----|------|-----|------|
| E | 1 | −1 | 1 | −1 |
| $C_2(z)$ | −1 | 1 | −1 | 1 |
| $\sigma_v(xz)$ | 1 | −1 | 1 | −1 |
| $\sigma_v'(yz)$ | −1 | 1 | −1 | 1 |

These $1 \times 1$ matrices therefore also constitute a *representation*, and this can be set out as

| E | $C_2(z)$ | $\sigma_v(xz)$ | $\sigma_v'(yz)$ |
|---|----------|----------------|-----------------|
| 1 | −1 | 1 | −1 |

(it is usual to ignore the '+' sign).

In the same way, if we consider the effect of the various symmetry operations on vectors $y_1$ and $z_1$, we arrive at two more representations. From $y_1$ we have (Fig. 2.8)

$$E(y_1) \rightarrow (y_1), \quad C_2(z)(y_1) \rightarrow -(y_1),$$

$$\sigma_v(xz)(y_1) \rightarrow -(y_1) \quad \sigma_v'(yz)(y_1) \rightarrow (y_1)$$

This produces the representation

| E | $C_2(z)$ | $\sigma_v(xz)$ | $\sigma_v'(yz)$ |
|---|----------|----------------|-----------------|
| 1 | −1 | −1 | 1 |

**Fig. 2.8**

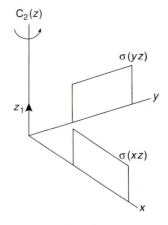

whilst from $z_1$ we have (Fig. 2.9)

| E | $C_2(z)$ | $\sigma_v(xz)$ | $\sigma_v'(yz)$ |
|---|----------|----------------|-----------------|
| 1 | 1 | 1 | 1 |

Both of these representations also obey the multiplication table.

**Fig. 2.9**

## 2.5    Irreducible representations: a second look at character tables

The three representations above have arisen as a result of using the vectors $x_1$, $y_1$ and $z_1$ to illustrate the effect of the various symmetry operations in $C_{2v}$. These representations are three of the *simplest representations* which exist for this point group, and are known as *irreducible representations*.

The vectors which led us to these irreducible representations are examples of bases, and we would say that here $x_1$, $y_1$ and $z_1$ are bases for the irreducible representations:

$$1 \quad -1 \quad 1 \quad -1; \quad 1 \quad -1 \quad -1 \quad 1; \quad \text{and} \quad 1 \quad 1 \quad 1 \quad 1$$

The fourth and final irreducible representation in $C_{2v}$ arises from the set of one-dimensional matrices:

$$E \quad C_2(z) \quad \sigma_v(xz) \quad \sigma_v'(yz)$$
$$1 \quad 1 \quad -1 \quad -1$$

### Representations in the $C_{2v}$ character table

In Chapter 1, character tables were introduced as providing a summary of the various symmetry operations in a point group, and we are now in a position to explore much of the additional information contained in these tables. We also know that the term *character* refers to the sum of the diagonal terms of a matrix. The complete character table for the point group $C_{2v}$ is reproduced below:

| $C_{2v}$ | $E$ | $C_2(z)$ | $\sigma_v(xz)$ | $\sigma_v'(yz)$ | $h = 4$ | |
|---|---|---|---|---|---|---|
| $A_1$ | 1 | 1 | 1 | 1 | $z$ | $x^2, y^2, z^2$ |
| $A_2$ | 1 | 1 | $-1$ | $-1$ | $R_z$ | $xy$ |
| $B_1$ | 1 | $-1$ | 1 | $-1$ | $x, R_y$ | $xz$ |
| $B_2$ | 1 | $-1$ | $-1$ | 1 | $y, R_x$ | $yz$ |

Along the top outer row are to be found the four symmetry operations in the group: the identity, the $C_2$ axis and the two planes. These symbols also confirm the location of the Cartesian axes. The main body of the table contains the *characters* of the *irreducible representations* discussed above, and because these representations are only one-dimensional matrices, the characters are identical to the matrices themselves.

The letters down the left-hand column are used to identify each representation by a specific label. The totally symmetric irreducible representation is labelled $A_1$, and refers to the row of $+1$ entries under each symmetry operation. Below this are the three other irreducible representations, labelled $A_2$, $B_1$ and $B_2$.

On the right-hand side of the table are to be found a number of mathematical functions, such as '$x$', or '$xy$', or related symbols such as '$R_x$' or '$R_y$'. There are usually two columns of these functions: the far right column listing squared or cross-product functions such as '$x^2$', '$y^2$' or '$xy$', and the inner column containing '$x$', '$y$' and '$z$', together with symbols denoting rotation about a particular axis—such as '$R_x$'.

These functions or symbols are placed on the same row as the irreducible representation to which they are related, and may be used as bases to illustrate the corresponding representations.

In Fig. 2.7, we saw the effect of the various symmetry operations in this point group on the vector $x_1$, and that the end result was to produce an irreducible representation:

$$E \quad C_2(z) \quad \sigma_v(xz) \quad \sigma_v'(yz)$$
$$1 \quad -1 \quad 1 \quad -1$$

This representation carries the label $B_1$ in the character table, and its relationship with the $x_1$ vector is indicated by placing the function '$x$' on the same row as $B_1$:

$$E \quad C_2(z) \quad \sigma_v(xz) \quad \sigma_v'(yz)$$
$$B_1 \quad 1 \quad -1 \quad 1 \quad -1 \quad x$$

There are various ways of expressing this correspondence between a mathematical function and its irreducible representation. One way is to say that '$x$ has the same symmetry as $B_1$', or '$x$ has $B_1$ symmetry', whilst another would be to say that '$x$ is a *basis* for the $B_1$ representation'. In the same way, the functions '$y$' and '$z$' in this column are linked to the $B_2$ and $A_1$ representations. In order to find a related function on which to base the $A_2$ representation, we enter the far right column. Here are to be found the squared and cross-product terms, and the character table indicates that one of these, '$xy$', could be used to illustrate the $A_2$ representation.

Figure 2.10 shows a $d_{xy}$ orbital in relation to the Cartesian $xy$ plane. The $z$ axis is perpendicular to the page. The wave function for this orbital has the same symmetry properties as the function '$xy$', in that it is positive when $x$ and $y$ are either both positive or both negative, and negative when $x$ and $y$ have opposite signs.

Here, the effect of carrying out the four symmetry operations is to produce the following relationships:

$$E(d_{xy}) \rightarrow (+1)(d_{xy}), \quad C_2(z)(d_{xy}) \rightarrow (+1)(d_{xy})$$
$$\sigma_v(xz)(d_{xy}) \rightarrow (-1)(d_{xy}), \quad \sigma_v'(yz)(d_{xy}) \rightarrow (-1)(d_{xy})$$

and the coefficients correspond to the $A_2$ representation.

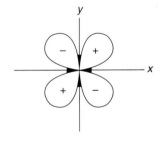

**Fig. 2.10**

### Irreducible representations in the $D_2$ character table

The multiplication table for $D_2$ which was derived earlier shows many similarities to that for $C_{2v}$, and these similarities persist when we compare the two character tables:

| $D_2$ | E | $C_2(z)$ | $C_2(y)$ | $C_2(x)$ | $h = 4$ | |
|-------|---|----------|----------|----------|---------|---|
| A | 1 | 1 | 1 | 1 | | $x^2, y^2, z^2$ |
| $B_1$ | 1 | 1 | $-1$ | $-1$ | $z, R_z$ | $xy$ |
| $B_2$ | 1 | $-1$ | 1 | $-1$ | $y, R_y$ | $xz$ |
| $B_3$ | 1 | $-1$ | $-1$ | 1 | $x, R_x$ | $yz$ |

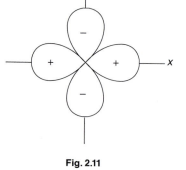

**Fig. 2.11**

The $D_2$ character table also has four irreducible representations, which can be written as one-dimensional matrices, and the functions $x$, $y$ and $z$ are again suitable bases for three of these—$B_3$, $B_2$ and $B_1$ respectively. In $D_2$ however, it is the totally symmetric representation, denoted here as 'A', which requires a basis function in the far right-hand column. This could be any of the terms $x^2$, $y^2$ or $z^2$, *or a sum or difference of these terms.*

Figure 2.11 shows the characteristic shape of the $d_{x^2-y^2}$ orbital in relation to the three $C_2$ axes in this symmetry group. Rotation about any of these $x$, $y$ or $z$ axes produces an equivalent configuration. All the equations describing the effect of the various symmetry elements on the $d_{x^2-y^2}$ orbital therefore involve the one-dimensional matrix ($+1$). This orbital could therefore be used as a basis for illustrating the 'A' representation in the $D_2$ point group.

## 2.6 Summary

This chapter has provided an introduction to the properties of simple matrices, and to their use as representations for symmetry operations. The idea that such operations can form a *group* is illustrated by the group multiplication table, and, in particular, it is shown how simple mathematical functions can act as bases for the irreducible representations in the point groups $C_{2v}$ and $D_2$. The exercises below provide an opportunity to become more familiar with the use of matrices and to further explore the use of character tables.

## 2.7 Exercises

$$P = \begin{pmatrix} 1 & 2 \\ 3 & 1 \end{pmatrix}$$

1. The matrices **P**, **Q** and **R** are reproduced alongside. Work out the *characters* of the matrices obtained as a result of the multiplications

$$Q = \begin{pmatrix} 3 & 1 \\ -2 & 0 \end{pmatrix}$$

(i) $P \times P$, (ii) $Q \times P$, (iii) $Q \times R$, (iv) $R \times P$

2. Using the matrices for the operations (E, $C_2(z)$, etc.) which are given in the text above, work out the *single* symmetry operations which are equivalent to (i) $C_2(x) \times \sigma(xy)$, (ii) $\sigma(xz) \times E \times i$, (iii) $C_2(z) \times C_2(x) \times \sigma(yz)$.

$$R = \begin{pmatrix} -1 & 1 \\ 7 & 3 \end{pmatrix}$$

3. Identify the representations to which functions $x$, $y^2$ and $xz$ respectively belong in the $D_{2h}$ point group (Appendix II).

4. Suggest which mathematical function(s) could be used as a basis for: (i) the $A_g$ representation in $D_{2h}$, (ii) the $A_u$ representation in $C_{2h}$, and (iii) the $B_{1g}$ representation in $D_{2h}$.

# 3 More on representations—the reduction formula

In Chapter 2, we explored the effect of symmetry operations such as a $C_2(z)$ rotation or a reflection $\sigma(xz)$ on simple vectors, and saw how this could lead to the concept of bases and irreducible representations in the $C_{2v}$ and $D_2$ point groups.

Irreducible representations were defined as the simplest representations which obeyed the group multiplication table, and for these point groups, these representations turned out to be rows of $1 \times 1$ matrices—effectively numbers—which could be identified by labels such as $A_1$, $B_2$, etc. The $C_{2v}$ multiplication table is reproduced below:

| $C_{2v}$ | E | $C_2(z)$ | $\sigma_v(xz)$ | $\sigma_v'(yz)$ |
|---|---|---|---|---|
| E | E | $C_2(z)$ | $\sigma_v(xz)$ | $\sigma_v'(yz)$ |
| $C_2(z)$ | $C_2(z)$ | E | $\sigma_v'(yz)$ | $\sigma_v(xz)$ |
| $\sigma_v(xz)$ | $\sigma_v(xz)$ | $\sigma_v'(yz)$ | E | $C_2(z)$ |
| $\sigma_v'(yz)$ | $\sigma_v'(yz)$ | $\sigma_v(xz)$ | $C_2(z)$ | E |

In this table, substitution of the symmetry operations E, $C_2(z)$, $\sigma(xz)$ and $\sigma(yz)$ by the $1 \times 1$ matrices, 1, 1, 1, and 1 respectively, retains the truth of the table, and these four matrices are labelled as the $A_1$ irreducible representation. Three other irreducible representations exist in this point group, and these have been labelled as $A_2$, $B_1$ and $B_2$.

## 3.1 Reducible representations

$$\begin{pmatrix} 1 & 0 \\ 0 & 1 \end{pmatrix}$$

**Fig. 3.1**

The sets of 'numbers' which define these four representations are the only such numbers which obey this multiplication table apart from a set of zeros, but it is not difficult to find *sets of matrices* which will do so. If we consider the $2 \times 2$ matrix shown in Fig. 3.1, then from the rules of matrix multiplication described earlier, we can show that

$$\begin{pmatrix} 1 & 0 \\ 0 & 1 \end{pmatrix} \times \begin{pmatrix} 1 & 0 \\ 0 & 1 \end{pmatrix} = \begin{pmatrix} 1 & 0 \\ 0 & 1 \end{pmatrix}$$

and, as a result, this matrix can take the place of each of the symmetry operations E, $C_2(z)$, $\sigma_v(xz)$ and $\sigma_v'(yz)$ in the $C_{2v}$ multiplication table. Four such matrices can therefore constitute a representation, in exactly the same way as did the 'numbers' 1, 1, 1 and 1.

However, the representation formed by these matrices is now *reducible*, rather than irreducible, as it can be broken down into two simpler representations.

Figure 3.2 shows how the four $2 \times 2$ matrices indicated above arise as a representation if we decide to carry out the symmetry operations of the $C_{2v}$ point group on the vectors $z_1$ and $v_1$. We have already seen (Fig. 2.9) that $z_1$ by *itself* can be used to illustrate the $A_1$ representation, and the same is clearly true for $v_1$, since it is also unaffected by any of the symmetry operations in the group. When $z_1$ and $v_1$ are considered *together*, the effect can be expressed, for example, as

$$E\begin{pmatrix} z_1 \\ v_1 \end{pmatrix} = \begin{pmatrix} 1 & 0 \\ 0 & 1 \end{pmatrix}\begin{pmatrix} z_1 \\ v_1 \end{pmatrix}$$

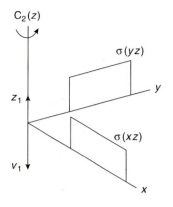

**Fig. 3.2**

The resulting representation, which can be labelled as '$\Gamma_{zv}$', now looks like this:

$$\begin{array}{cccc} E & C_2(z) & \sigma_v(xz) & \sigma_v'(yz) \\ \begin{pmatrix} 1 & 0 \\ 0 & 1 \end{pmatrix} & \begin{pmatrix} 1 & 0 \\ 0 & 1 \end{pmatrix} & \begin{pmatrix} 1 & 0 \\ 0 & 1 \end{pmatrix} & \begin{pmatrix} 1 & 0 \\ 0 & 1 \end{pmatrix} \end{array}$$

This representation is clearly made up of the $A_1$ representation for $z_1$ and the $A_1$ representation for $v_1$, and the presence of these two components is illustrated below:

$$\begin{pmatrix} z_1 \\ v_1 \end{pmatrix}\begin{pmatrix} 1 & 0 \\ 0 & 1 \end{pmatrix}\begin{pmatrix} 1 & 0 \\ 0 & 1 \end{pmatrix}\begin{pmatrix} 1 & 0 \\ 0 & 1 \end{pmatrix}\begin{pmatrix} 1 & 0 \\ 0 & 1 \end{pmatrix} \quad \begin{array}{c} A_1 \\ \\ A_1 \end{array}$$

with the reduction step being summarised as $\Gamma_{zv} = 2A_1$.

Figure 3.3 sets up the procedure for identifying the reducible representation $\Gamma_{xyz}$ which describes the result of using the three vectors $x_1$, $y_1$ and $z_1$ as bases. These matrices constitute the representation $\Gamma_{xyz}$.

The $3 \times 3$ matrices which now arise are identical to those derived previously in Chapter 2 when considering the effect of the various symmetry operations on the point $P(X,Y,Z)$, and are

$$\begin{array}{cccc} \begin{pmatrix} 1 & 0 & 0 \\ 0 & 1 & 0 \\ 0 & 0 & 1 \end{pmatrix} & \begin{pmatrix} -1 & 0 & 0 \\ 0 & -1 & 0 \\ 0 & 0 & 1 \end{pmatrix} & \begin{pmatrix} 1 & 0 & 0 \\ 0 & -1 & 0 \\ 0 & 0 & 1 \end{pmatrix} & \begin{pmatrix} -1 & 0 & 0 \\ 0 & 1 & 0 \\ 0 & 0 & 1 \end{pmatrix} \\ E & C_2(z) & \sigma(xz) & \sigma(yz) \end{array}$$

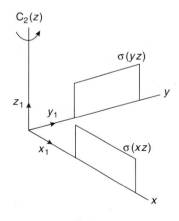

**Fig. 3.3**

Here, similar tie-lines reveal the irreducible representations to be

$$\Gamma_{xyz} = B_1 + B_2 + A_1$$

This is an important representation which relates both to atomic motion and to the symmetry of atomic orbitals.

In these two examples, it is clear that the results are unique, and that there are no other possible solutions for the reduction process. This turns out to be a general feature of reducible representations:

*A reducible representation gives rise to only one set of irreducible representations.*

## 3.2   Use of matrix characters in reduction by inspection

For the simple $2 \times 2$ and $3 \times 3$ matrices shown above, it is evident that the character of each matrix is the *sum* of the characters of the corresponding irreducible matrices, i.e.

$$\chi_R = \sum \chi_I$$

where $\chi_R$ and $\chi_I$ refer to the characters of the reducible and irreducible matrices respectively. This is a *quite general result*, and can sometimes be used to obtain irreducible representations by inspection.

For example, the representation $\Gamma_{zv}$ above has the characters: $\chi_R = 2 \ \ 2 \ \ 2 \ \ 2$ for the four matrices, and in the $C_{2v}$ point group this can only correspond to

$$
\begin{array}{ccccc}
 & E & C_2(z) & \sigma_v(xz) & \sigma_v'(yz) \\
 & 1 & 1 & 1 & 1 \\
+ & 1 & 1 & 1 & 1
\end{array}
$$

i.e. $2A_1$.

When the matrices which form the reducible representation are small — e.g. for $2 \times 2$ matrices, and perhaps also for $3 \times 3$ matrices — this method of *reduction by inspection* is one of the quickest ways of obtaining the irreducible representations, and it can always be used to *check* that the correct irreducible representations have been derived from a given set of characters.

*Example.* In the point group $C_{2v}$ show that a representation $\Gamma_1$ which consists of matrices with the characters 5 3 1 −1 can be reduced to $2A_1 + 2A_2 + B_1$.

Here, the appropriate check can be made by summing the columns for each operation in the $C_{2v}$ character table

| $C_{2v}$ | E | $C_2$ | $\sigma_v(xz)$ | $\sigma_v'(yz)$ | $h = 4$ | |
|------|---|-------|------|------|---|-----|
| $A_1$ | 1 | 1 | 1 | 1 | $z$ | $x^2, y^2, z^2$ |
| $A_2$ | 1 | 1 | −1 | −1 | $R_z$ | $xy$ |
| $B_1$ | 1 | −1 | 1 | −1 | $x, R_y$ | $xz$ |
| $B_2$ | 1 | −1 | −1 | 1 | $y, R_x$ | $yz$ |

| | E | $C_2(z)$ | $\sigma_v(xz)$ | $\sigma_v'(yz)$ |
|------|---|-------|------|------|
| $A_1$ | 1 | 1 | 1 | 1 |
| $A_1$ | 1 | 1 | 1 | 1 |
| $A_2$ | 1 | 1 | −1 | −1 |
| $A_2$ | 1 | 1 | −1 | −1 |
| $B_1$ | 1 | −1 | 1 | −1 |
| $\Gamma_1$ | 5 | 3 | 1 | −1 |

## 3.3 Reduction of representations using the 'reduction formula'

In the example above, it is straightforward to *confirm* that a particular set of irreducible representations arise from the given characters of $\Gamma_1$, but to *derive* the five irreducible representations would initially seem to involve an almost hit-or-miss approach involving the arbitrary selection of rows of numbers until agreement is reached. Fortunately, there exists a general procedure for obtaining the irreducible representations which constitute *any* reducible representation. This procedure is contained in the *reduction formula*, and involves only the characters of the matrices of the reducible representation together with information contained in the relevant character table.

This formula sets out to identify the *number of each type of irreducible representation* contained in a *given reducible representation*, and is most conveniently defined and illustrated by reference to a specific example. The formula is generally given as

$$ n = \frac{1}{h} \sum N \chi_R \chi_I $$

Here, '$n$' is the number of times a particular irreducible representation occurs. $\chi_R$ and $\chi_I$ are the characters of the reducible and irreducible representations. $N$ is the coefficient in front of each of the symmetry element symbols listed on the top row of the character table. '$h$' is the *order* of the group, and is the sum of the coefficients of the symmetry element symbols, i.e. $h = \sum N$. It is also, as mentioned in Chapter 1, the maximum number of equivalent points generated from a single point by all the symmetry operations in the group.

The summation in the reduction formula is carried out over each of the columns in the character table for the irreducible representation under consideration. We shall apply the reduction formula first to the representation $\Gamma_1$ in the $C_{2v}$ point group:

| $C_{2v}$ | E | $C_2(z)$ | $\sigma_v(xz)$ | $\sigma_v'(yz)$ |
|----------|---|----------|----------------|-----------------|
| $\Gamma_1$ | 5 | 3 | 1 | $-1$ |

In the $C_{2v}$ point group, the coefficients $N$ of E, $C_2(z)$, $\sigma_v(xz)$ and $\sigma_v'(yz)$ are all 1, and the order of the group is 4.

| $C_{2v}$ | E | $C_2$ | $\sigma_v(xz)$ | $\sigma_v'(yz)$ | $h = 4$ | |
|----------|---|-------|----------------|-----------------|---------|---|
| $A_1$ | 1 | 1 | 1 | 1 | $z$ | $x^2, y^2, z^2$ |
| $A_2$ | 1 | 1 | $-1$ | $-1$ | $R_z$ | $xy$ |
| $B_1$ | 1 | $-1$ | 1 | $-1$ | $x, R_y$ | $xz$ |
| $B_2$ | 1 | $-1$ | $-1$ | 1 | $y, R_x$ | $yz$ |

The number of times a particular representation occurs is obtained by focusing on the appropriate row of characters for the representation, and first calculating the product $N\chi_R\chi_I$ for each of the symmetry elements in turn. Thus, in order to calculate the number of $A_1$ representations $n(A_1)$, we focus on the values of $\chi_I$ for this representation, which are $+1$ for each operation. We can then set up the listing as follows:

$$N \times \chi_R \times \chi_I$$

$$
\begin{array}{llll}
\text{for E} & 1 \times & 5 \times 1 = 5 \\
\text{for } C_2(z) & 1 \times & 3 \times 1 = 3 \\
\text{for } \sigma_v(xz) & 1 \times & 1 \times 1 = 1 \\
\text{for } \sigma'_v(yz) & 1 \times & -1 \times 1 = -1
\end{array}
$$

The summation of these terms, $\sum N\chi_R\chi_I$ is 8, and as the order of the group is 4, the number of $A_1$ representations is $n(A_1) = 2$.

The number of $A_2$ representations is similarly obtained by focusing on the $\chi_I$ values $1\ 1\ -1\ -1$ which relate to the $A_2$ representation in $C_{2v}$. Here, we may evaluate $n(A_2)$ as

$$
\begin{aligned}
n(A_2) &= (1/4)\{(1 \times 5 \times 1) + (1 \times 3 \times 1) \\
&\quad + (1 \times 1 \times -1) + (1 \times -1 \times -1)\} \\
&= (1/4)\{5 + 3 - 1 + 1\} = 2
\end{aligned}
$$

In a similar way,

$$
\begin{aligned}
n(B_1) &= (1/4)\{(1 \times 5 \times 1) + (1 \times 3 \times -1) + (1 \times 1 \times 1) \\
&\quad + (1 \times -1 \times -1)\} = 1
\end{aligned}
$$

This result is identical to that obtained previously, and to confirm that there are no $B_2$ representations contained within $\Gamma_1$, we can evaluate $n(B_2)$ as

$$n(B_2) = (1/4)\{5 - 3 - 1 - 1\} = 0$$

If the value of '$n$' obtained from the reduction formula is negative or fractional, then either a mistake has been made in the above arithmetic process, or one or more of the characters $\chi_R$ of the reducible representation is incorrect.

## 3.4   Widening the bases for representations

The representations we have considered so far have arisen by considering the effect of symmetry operations on specific points in space—e.g. on the point $P(X, Y, Z)$—or on vectors such as $x_1$, $y_1$, etc. However, the most obvious manifestation of symmetry in chemistry is the equivalence of atoms, orbitals, bonds and angles, and it is now appropriate to consider these objects also as bases for representations.

### Atoms, bonds and orbitals in the $H_2O$ molecule

Figure 3.4 shows the $H_2O$ molecule in relation to the symmetry elements of the $C_{2v}$ point group to which it belongs. The hydrogen atoms may be identified as $H_1$ and $H_2$, and the bonds which connect these atoms to oxygen are labelled $r_1$ and $r_2$. The effect of carrying out the $C_{2v}$ symmetry operations on the two hydrogen atoms is easy to visualise. The identity E, and reflection in the plane $\sigma_v(xz)$ leave both H atoms unshifted, but rotation by $C_2(z)$ and reflection in $\sigma_v'(yz)$ both cause $H_1$ and $H_2$ to change places. The matrices which express this result are then

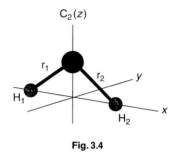

**Fig. 3.4**

$$\begin{array}{cccc} \text{E} & C_2(z) & \sigma_v(xz) & \sigma_v'(yz) \\ \begin{pmatrix} 1 & 0 \\ 0 & 1 \end{pmatrix} & \begin{pmatrix} 0 & 1 \\ 1 & 0 \end{pmatrix} & \begin{pmatrix} 1 & 0 \\ 0 & 1 \end{pmatrix} & \begin{pmatrix} 0 & 1 \\ 1 & 0 \end{pmatrix} \end{array}$$

These matrices form a representation which we can label $\Gamma_H$. The characters of the matrices are 2 0 2 0, and this representation reduces to

$$\Gamma_H = A_1 + B_1$$

If we now consider the effect of the symmetry operations on the pair of bonds $r_1$ and $r_2$, *exactly the same matrices emerge as for the two hydrogen atoms*. The operations E and $\sigma_v(xz)$ leave the bonds unshifted, resulting in matrices with characters of 2, whilst $C_2(z)$ and $\sigma_v'(yz)$ interchange the bonds, and give matrices with characters of 0. We can therefore write

$$\Gamma_r = A_1 + B_1$$

In a similar way, we can obtain the representation for the $1s$ *atomic orbitals* on the two hydrogen atoms. If the orbitals on $H_1$ and $H_2$ are identified as $\phi_1$ and $\phi_2$, then it can be seen that these orbitals transform in exactly the same way as the parent atoms:

$$\Gamma_\phi = A_1 + B_1$$

Finally, we should consider the effect of the symmetry operations on the oxygen atom and its orbitals.

   The oxygen atom in $H_2O$ lies on all the symmetry elements, and remains unshifted for all the symmetry operations. We may therefore write

$$\Gamma_0 = A_1$$

The effect of the symmetry operations on the oxygen $1s$, $2s$ and $2p$ *orbitals* is most conveniently explored by direct reference to the $C_{2v}$ character table, and to the functions which have the same symmetry properties as the $s$, $p_x$, $p_y$, and $p_z$ orbitals, using the procedure discussed previously in Chapter 2. The $1s$ and $2s$ orbitals on the oxygen atom have spherical symmetry, and so transform as $A_1$, whilst the $p_x$, $p_y$ and $p_z$ orbitals have the same symmetry as the functions $x$, $y$ and $z$ respectively.

The $C_{2v}$ character table identifies the representations for $x$, $y$ and $z$ as $B_1$, $B_2$ and $A_1$ respectively, and we may therefore summarise the symmetry properties of the oxygen orbitals by writing

$$\Gamma_{1s} = A_1, \qquad \Gamma_{2s} = A_1, \qquad \Gamma_{2p} = A_1 + B_1 + B_2$$

This last procedure—identifying the irreducible representations for sets of atomic orbitals—is of fundamental importance in molecular orbital theory, and the previous exercise involving the derivation of the representations for the two O–H bonds would be a typical prerequisite to establishing the number and activity of molecular stretching vibrations in $H_2O$ or to describing the $\sigma$-bonding scheme.

**Bonds in the $C_2H_4$ molecule**

As a final worked example of the use of the reduction formula, we shall work out the representations for the five $\sigma$-bonds in $C_2H_4$. Ethylene has a planar structure and belongs to the $D_{2h}$ point group, and the $D_{2h}$ character table is reproduced below:

| $D_{2h}$ | $E$ | $C_2(z)$ | $C_2(y)$ | $C_2(x)$ | $i$ | $\sigma(xy)$ | $\sigma(xz)$ | $\sigma(yz)$ | $h = 8$ | |
|---|---|---|---|---|---|---|---|---|---|---|
| $A_g$ | 1 | 1 | 1 | 1 | 1 | 1 | 1 | 1 | | $x^2, y^2, z^2$ |
| $B_{1g}$ | 1 | 1 | $-1$ | $-1$ | 1 | 1 | $-1$ | $-1$ | $R_z$ | $xy$ |
| $B_{2g}$ | 1 | $-1$ | 1 | $-1$ | 1 | $-1$ | 1 | $-1$ | $R_y$ | $xy$ |
| $B_{3g}$ | 1 | $-1$ | $-1$ | 1 | 1 | $-1$ | $-1$ | 1 | $R_x$ | $yz$ |
| $A_u$ | 1 | 1 | 1 | 1 | $-1$ | $-1$ | $-1$ | $-1$ | | |
| $B_{1u}$ | 1 | 1 | $-1$ | $-1$ | $-1$ | $-1$ | 1 | 1 | $z$ | |
| $B_{2u}$ | 1 | $-1$ | 1 | $-1$ | $-1$ | 1 | $-1$ | 1 | $y$ | |
| $B_{3u}$ | 1 | $-1$ | $-1$ | 1 | $-1$ | 1 | 1 | $-1$ | $x$ | |

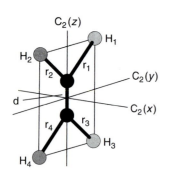

**Fig. 3.5**

As noted earlier, this point group has three mutually perpendicular $C_2$ axes, and the orientation of a molecule with respect to these axes is to some extent arbitrary. In the case of $C_2H_4$, it is convenient to choose the direction of the C–C bond as $z$, and $x$ to be perpendicular to the plane of the molecule. Figure 3.5 shows the $C_2H_4$ molecule within this coordinate system.

The four C–H bonds are equivalent, and may be labelled $r_1, \ldots, r_4$, and there is one C–C bond labelled 'd'. The four hydrogen atoms are similarly equivalent, and are labelled $H_1$ to $H_4$ as shown.

## 3.5   The representations $\Gamma_r$ and $\Gamma_d$

In order to derive the irreducible representations for $\Gamma_r$, we need first to establish the characters $\chi_R$ of the reducible representation. This could be done by working out the complete matrices which describe the effect of the various symmetry operations on the four bonds $r_1, \ldots, r_4$, and then summing the terms on each diagonal. This would be a lengthy process and, fortunately, there is a much more rapid method of obtaining these characters.

In the $H_2O$ example above, we saw that the operations E and $\sigma(xz)$ left the two O–H bonds unshifted and that, as a result, each bond was found to contribute +1 to the character. In contrast, the operations $C_2$ and $\sigma(yz)$ moved both bonds, leaving zeros on the diagonal. In order to establish the values of $\chi_R$ for the various matrices, it was therefore necessary only to consider the (+1) contribution from any *unshifted bonds*.

In the case of $C_2H_4$, the identity E leaves all four bonds in their *original positions*, and so the character for the matrix describing the effect of E has the value $\chi_R = 4$. In a similar way, the character for reflection in the plane of the molecule $\sigma(yz)$ is also 4, since no bond is shifted.

For each of the other operations *every bond is moved*, and so the characters for all these matrices are zero.

Using this simplification, the characters $\chi_R$ for $\Gamma_r$ are

| $D_{2h}$ | E | $C_2(z)$ | $C_2(y)$ | $C_2(x)$ | i | $\sigma(xy)$ | $\sigma(xz)$ | $\sigma(yz)$ |
|---|---|---|---|---|---|---|---|---|
| $\Gamma_r$ | 4 | 0 | 0 | 0 | 0 | 0 | 0 | 4 |

If we now apply the reduction formula

$$n = \frac{1}{h}\sum N\chi_R\chi_I$$

we can see that the presence of the zeros in $\chi_R$ means that in the summation we need only consider *two* terms: those for the identity E and for $\sigma(yz)$. All the other terms will be zero.

The coefficients $N$ are all +1, and the order of the $D_{2h}$ point group is 8. Values of '$n$' may then be obtained as follows:

$$n(A_g) = (1/8)\{(1 \times 4 \times 1) + (1 \times 4 \times 1)\} = 1$$
$$n(B_{1g}) = (1/8)\{(1 \times 4 \times 1) - (1 \times 4 \times 1)\} = 0$$
$$n(B_{2g}) = (1/8)\{(1 \times 4 \times 1) - (1 \times 4 \times 1)\} = 0$$
$$n(B_{3g}) = (1/8)\{(1 \times 4 \times 1) + (1 \times 4 \times 1)\} = 1$$
$$n(A_u) = (1/8)\{(1 \times 4 \times 1) - (1 \times 4 \times 1)\} = 0$$
$$n(B_{1u}) = (1/8)\{(1 \times 4 \times 1) + (1 \times 4 \times 1)\} = 1$$
$$n(B_{2u}) = (1/8)\{(1 \times 4 \times 1) + (1 \times 4 \times 1)\} = 1$$
$$n(B_{3u}) = (1/8)\{(1 \times 4 \times 1) - (1 \times 4 \times 1)\} = 0$$

The representation $\Gamma_r$ therefore reduces to

$$\Gamma_r = A_g + B_{3g} + B_{1u} + B_{2u}$$

The representation $\Gamma_d$ is obtained very easily. There is only one bond 'd', and it is left unchanged by all the symmetry operations. Each operation is therefore represented by the $1 \times 1$ matrix $(+1)$, which corresponds to the totally symmetric representation $A_g$:

$$\Gamma_d = A_g$$

The representations obtained above for $\Gamma_r$ and $\Gamma_d$ illustrate a general result which arises in connection with representations based on bonds.

For any set of equivalent bonds, it can be shown that the *irreducible representation for these bonds* **must contain** *the totally symmetric representation*.

The corollary to this rule is that *if there is only one bond of a particular type in a molecule, this bond must transform according to the totally symmetric representation*.

## 3.6   Summary

The main purpose of this chapter has been to explore the use of matrices as *reducible* representations, and to show how these lead to *irreducible* representations by use of the reduction formula. At the same time, the concept of using other bases for representations, such as atoms, bonds and orbitals, is introduced.

## 3.7   Exercises

1.   Reduce the representations $\Gamma_1$ and $\Gamma_2$ below in the $C_{2v}$ point group:

| $C_{2v}$ | E | $C_2(z)$ | $\sigma_v(xz)$ | $\sigma_v'(yz)$ |
|---|---|---|---|---|
| $\Gamma_1$ | 4 | 2 | 0 | -2 |
| $\Gamma_2$ | 4 | 0 | 0 | 0 |

2   Derive the representation $\Gamma_{C-H}$ for the following molecules:
(i) *trans* CHCl=CHCl, (ii) $C_6H_5Cl$: assume the molecule lies in the $xz$ plane; (iii) $1,4\text{-}C_6H_4Cl_2$: assume the molecule lies in the $yz$ plane.

# 4 Matrices and representations in higher order point groups—degenerate representations

In the previous two chapters, we have encountered matrices which describe the operations E, i, $\sigma$ and $C_2$, and although some of these matrices have been of relatively high order, up to $3 \times 3$, it has always been possible to simplify the corresponding reducible representations so that the final irreducible representations have been $1 \times 1$ matrices—effectively 'numbers'.

This situation no longer applies when we consider higher order rotation axes, and the higher order point groups which contain them.

## 4.1 Matrices for $C_4$ rotations

Figure 4.1 provides a framework for *contrasting* the effect of a $C_2(z)$ rotation on a typical point $P(X, Y)$ in the $xy$ plane, with a $C_4$ rotation about $z$. Rotation through $180°$ in an anticlockwise direction takes P to a new position $P'$ with coordinates $X'$, $Y'$ such that

$$X' = -X \quad \text{and} \quad Y' = -Y$$

The single matrix equation which describes these operations is

$$C_2(z)\begin{pmatrix} X \\ Y \end{pmatrix} = \begin{pmatrix} -1 & 0 \\ 0 & -1 \end{pmatrix}\begin{pmatrix} X \\ Y \end{pmatrix}$$

but it can clearly be simplified to give two equations involving only $1 \times 1$ matrices.

$$C_2(z)(X) = (-1)(X) \quad \text{and} \quad C_2(z)(Y) = (-1)(Y)$$

In contrast, rotation through $90°$ in an anticlockwise direction takes P to the position $P''$ with coordinates $X''$, $Y''$ such that

$$X'' = -Y \quad \text{and} \quad Y'' = X$$

with a corresponding matrix equation

$$C_4^1(z)\begin{pmatrix} X \\ Y \end{pmatrix} = \begin{pmatrix} 0 & -1 \\ 1 & 0 \end{pmatrix}\begin{pmatrix} X \\ Y \end{pmatrix}$$

However, this matrix for $C_4^1(z)$ *cannot be further simplified*. The values of $X$ and $Y$ are inextricably scrambled by the $C_4$ operation. Rotation through

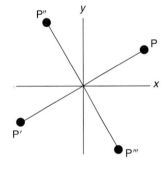

Fig. 4.1

270°, which corresponds to $C_4^3$, takes the point $P(X, Y)$ to the position $P'''$ with coordinate values $(Y, -X)$, and the matrix which describes this is

$$C_4^3(z)\begin{pmatrix} X \\ Y \end{pmatrix} = \begin{pmatrix} 0 & 1 \\ -1 & 0 \end{pmatrix}\begin{pmatrix} X \\ Y \end{pmatrix}$$

This matrix similarly scrambles $X$ and $Y$ and although it is different from the matrix for $C_4^1$, it is significant that the *characters* of the $C_4^1$ and $C_4^3$ *matrices are the same*.

## 4.2   Matrices for rotation through an angle $\theta$: a general expression for the character of the $C_n^1$ operation

The effect of an anticlockwise rotation about the $z$ axis through a general angle $\theta$ is illustrated in Fig. 4.2, where a point $Q(X, Y)$ lying in the $xy$ plane moves to a position $Q'(X', Y')$. If we identify the distance of Q and Q' from the coordinate origin as '$d$', then the initial coordinates of $Q(X, Y)$ may be alternatively expressed as

$$X = d\cos\alpha, \qquad Y = d\sin\alpha$$

After rotation, the new coordinates $(X', Y')$ are given by

$$X' = d\cos(\alpha + \theta), \quad Y' = d\sin(\alpha + \theta)$$

Expansion of these standard trigonometric expressions gives

$$X' = d\{\cos\alpha\cos\theta - \sin\alpha\sin\theta\} = X\cos\theta - Y\sin\theta$$
$$Y' = d\{\sin\alpha\cos\theta + \cos\alpha\sin\theta\} = Y\cos\theta + X\sin\theta$$

Rotation about $z$ through the angle $\theta$ $(n = 360/\theta)$ in an anticlockwise direction can therefore be summarised by the matrix equation

$$C_n^1\begin{pmatrix} X \\ Y \end{pmatrix} = \begin{pmatrix} \cos\theta & -\sin\theta \\ \sin\theta & \cos\theta \end{pmatrix}\begin{pmatrix} X \\ Y \end{pmatrix}$$

The character of this matrix is therefore $2\cos\theta$.

In general, point Q will also have a $z$ coordinate $Z$, and the action of $C_n^1$ will move $Q(X, Y, Z)$ to $Q'(X', Y', Z')$. However, as the $C_n$ axis lies along the $z$ axis, the effect of this rotation on the coordinate $Z$ is to leave it unchanged:

$$Z' = Z, \text{ i.e. } C_n^1(Z) = (+1)(Z)$$

If we *incorporate* this addition into our description of this operation, the matrix which describes the effect of $C_n^1$ on a *general point* $(X, Y, Z)$ can be seen to be

$$C_n^1\begin{pmatrix} X \\ Y \\ Z \end{pmatrix} = \begin{pmatrix} \cos\theta & -\sin\theta & 0 \\ \sin\theta & \cos\theta & 0 \\ 0 & 0 & 1 \end{pmatrix}\begin{pmatrix} X \\ Y \\ Z \end{pmatrix}$$

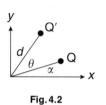

**Fig. 4.2**

and this has the character

$$\chi C_n = 2\cos\theta + 1$$

This is an important general result, as it leads directly to the character of the matrix for rotation about an axis of any order.

This equation also yields a further general result concerning the characters of $C_n^m$ matrices:

*The character of the matrix for the $C_n^1$ operation will be identical to the character of the matrix for $C_n^{-1}$ (i.e. $C_n^{n-1}$).*

This equivalence arises because $\cos\theta = \cos(-\theta)$, and it has the effect that the character is *independent* both of the '*direction*' of the axis and of the *sense of rotation*.

It therefore sidesteps any ambiguity regarding both the 'direction' of an axis (relative to atomic positions in a molecule) and the subsequent direction of rotation (clockwise or anticlockwise) when higher order axes are involved. Both these directions are sometimes treated in a somewhat arbitrary way without prior qualification.

## 4.3  Representations in higher order point groups

We are now in a position to derive representations in the majority of point groups, using vectors, atoms, bonds or orbitals as bases, and this chapter is mainly devoted to illustrating these procedures.

Before reaching this stage, however, it is useful to become better acquainted with the symmetry properties and character tables for two of the most commonly encountered point groups, $C_{3v}$ and $T_d$.

## 4.4  Representations in the $C_{3v}$ point group

The $C_{3v}$ point group contains three kinds of symmetry elements: E, $C_3$ and $\sigma_v$. The top row of the character table acknowledges the two operations associated with the $C_3$ axis, $C_3^1$ and $C_3^2$, by the coefficient $N=2$, and the three vertical planes appear as a single entry $3\sigma_v$:

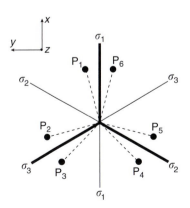

| $C_{3v}$ | E | $2C_3$ | $3\sigma_v$ | $h=6$ | |
|---|---|---|---|---|---|
| $A_1$ | 1 | 1 | 1 | $z$ | $x^2+y^2, z^2$ |
| $A_2$ | 1 | 1 | −1 | $R_z$ | |
| E | 2 | −1 | 0 | $(x,y),$ | $(x^2-y^2, xy),$ |
| | | | | $(R_x, R_y)$ | $(xz, yz)$ |

The order of the group, $h$, is 6, and Fig. 4.3 shows a view down the $C_3$ axis depicting the six points $P_1$ to $P_6$ which can be generated from an original point $P_1$ by the axis and planes present.

**Fig. 4.3**

The effect of the symmetry operations on $P_1$ may be summarised as

$$E(P_1) = P_1, \qquad C_3^1(P_1) = P_3, \qquad C_3^2(P_1) = P_5,$$
$$\sigma_1(P_1) = P_6, \qquad \sigma_2(P_1) = P_2, \qquad \sigma_3(P_1) = P_4$$

and the group multiplication table takes the form.

| $C_{3v}$ | E | $C_3^1$ | $C_3^2$ | $\sigma_1$ | $\sigma_2$ | $\sigma_3$ |
|---|---|---|---|---|---|---|
| E | E | $C_3^1$ | $C_3^2$ | $\sigma_1$ | $\sigma_2$ | $\sigma_3$ |
| $C_3^1$ | $C_3^1$ | $C_3^2$ | E | $\sigma_2$ | $\sigma_3$ | $\sigma_1$ |
| $C_3^2$ | $C_3^2$ | E | $C_3^1$ | $\sigma_3$ | $\sigma_1$ | $\sigma_2$ |
| $\sigma_1$ | $\sigma_1$ | $\sigma_3$ | $\sigma_2$ | E | $C_3^2$ | $C_3^1$ |
| $\sigma_2$ | $\sigma_2$ | $\sigma_1$ | $\sigma_3$ | $C_3^1$ | E | $C_3^2$ |
| $\sigma_3$ | $\sigma_3$ | $\sigma_2$ | $\sigma_1$ | $C_3^2$ | $C_3^1$ | E |

In this multiplication table, it may be noted that some sequential symmetry operations are not commutative, e.g.

$$\sigma_1 \times C_3^1 = \sigma_3, \quad \text{but} \quad C_3^1 \times \sigma_1 = \sigma_2$$

The left-hand column of the character table shows three irreducible representations, $A_1$, $A_2$ and E, and the final columns indicate functions that could be used as appropriate bases.

The first two representations look similar to those previously encountered in the point groups $C_{2v}$. They have labels which are familiar, and one can imagine that the rows of numbers might similarly correspond to $1 \times 1$ matrices. The character table identifies $z$ as a function with $A_1$ symmetry, and this may be confirmed by noting that a vector $z_1$ located at the coordinate origin (Fig. 4.3) would be unshifted by any of the symmetry operations, as it lies on the $C_3$ axis and in all three planes.

The representation labelled 'E', however, looks different. Firstly, the use of this symbol for a representation is new, and the row of numbers contains a 2 and a 0, both of which are unfamiliar. In addition, the functions which might be used to illustrate this representation now appear as *pairs* within a bracket, e.g. $(x, y)$ or $(xz, yz)$, rather than singly.

## 4.5 Representations based on *X*, *Y* and *Z* in the $C_{3v}$ point group

E

$$\begin{pmatrix} 1 & 0 & 0 \\ 0 & 1 & 0 \\ 0 & 0 & 1 \end{pmatrix}$$

In order to understand these entries in the $C_{3v}$ character table, it is useful to derive the matrices for the various symmetry operations using the coordinates of point $P_1$ in Fig. 4.3. If these are taken to be $(X, Y, Z)$, then the matrices for four of the six operations can be readily obtained. For the identity E, $P_1$ remains unshifted, and the matrix is shown alongside.

The matrices for the operations $C_3^1$ and $C_3^2$ follow directly from the general expression

$$C_n^i \begin{pmatrix} X \\ Y \\ Z \end{pmatrix} = \begin{pmatrix} \cos\theta & -\sin\theta & 0 \\ \sin\theta & \cos\theta & 0 \\ 0 & 0 & 1 \end{pmatrix} \begin{pmatrix} X \\ Y \\ Z \end{pmatrix}$$

Substituting $\theta$ values of 120° and 240° respectively, we have

$$C_3^1 \qquad\qquad\qquad C_3^2$$

$$\begin{pmatrix} -1/2 & -(\sqrt{3})/2 & 0 \\ (\sqrt{3})/2 & -1/2 & 0 \\ 0 & 0 & 1 \end{pmatrix} \qquad \begin{pmatrix} -1/2 & (\sqrt{3})/2 & 0 \\ -(\sqrt{3})/2 & -1/2 & 0 \\ 0 & 0 & 1 \end{pmatrix}$$

These two matrices are therefore *different*, but they have *the same character*, in this case, zero. Reflection in $\sigma_1$ takes $P_1(X, Y, Z)$ to $P_6(X, -Y, Z)$, and this matrix is therefore

$$\sigma_1 = \begin{pmatrix} 1 & 0 & 0 \\ 0 & -1 & 0 \\ 0 & 0 & 1 \end{pmatrix}$$

with a character $\chi = 1$.

The matrices for reflection in the planes $\sigma_2$ and $\sigma_3$ are more complicated, but it can be shown that for *any vertical plane* oriented at an angle $\phi$ to the $x$ axis, the general matrix for reflection of the point $P(X, Y, Z)$ in the plane takes the form

$$\sigma \begin{pmatrix} X \\ Y \\ Z \end{pmatrix} = \begin{pmatrix} \cos 2\phi & \sin 2\phi & 0 \\ \sin 2\phi & -\cos 2\phi & 0 \\ 0 & 0 & 1 \end{pmatrix}$$

The character of this matrix is *independent* of the value of $\phi$, as the $\cos 2\phi$ terms on the diagonal cancel, giving a net $\chi = 1$. For $\sigma_2$ and $\sigma_3$ in $C_{3v}$ symmetry, $\phi$ takes the values 120° and 240° respectively, and the six $3 \times 3$ matrices may therefore be identified as follows:

$$E \qquad\qquad\qquad C_3^1 \qquad\qquad\qquad C_3^2$$

$$\begin{pmatrix} 1 & 0 & 0 \\ 0 & 1 & 0 \\ 0 & 0 & 1 \end{pmatrix} \quad \begin{pmatrix} -1/2 & -(\sqrt{3})/2 & 0 \\ (\sqrt{3})/2 & -1/2 & 0 \\ 0 & 0 & 1 \end{pmatrix} \quad \begin{pmatrix} -1/2 & (\sqrt{3})/2 & 0 \\ -(\sqrt{3})/2 & -1/2 & 0 \\ 0 & 0 & 1 \end{pmatrix}$$

$$\sigma_1 \qquad\qquad\qquad \sigma_2 \qquad\qquad\qquad \sigma_3$$

$$\begin{pmatrix} 1 & 0 & 0 \\ 0 & -1 & 0 \\ 0 & 0 & 1 \end{pmatrix} \quad \begin{pmatrix} -1/2 & -(\sqrt{3})/2 & 0 \\ -(\sqrt{3})/2 & 1/2 & 0 \\ 0 & 0 & 1 \end{pmatrix} \quad \begin{pmatrix} -1/2 & (\sqrt{3})/2 & 0 \\ (\sqrt{3})/2 & 1/2 & 0 \\ 0 & 0 & 1 \end{pmatrix}$$

It can be seen that these can be simplified into a *set of 2 × 2 matrices linked to X and Y*, and a $1 \times 1$ matrix linked to Z. These $2 \times 2$ matrices arise in a very similar way to the $2 \times 2$ matrices found earlier for the $C_4^1$ operation. Here in the $C_{3v}$ point group, the $C_3$ operations again scramble X and Y, with the result that they cannot be separated, and when taken *together*, they form the basis for a *doubly degenerate irreducible representation*.

## 4.6   Irreducible representations from the matrices based on *X*, *Y* and *Z* in the C₃ᵥ point group

At this stage, we can identify the characters $\chi_R$ of the six matrices describing *all the operations* as follows:

| $C_{3v}$ | E | $C_3^1$ | $C_3^2$ | $\sigma_1$ | $\sigma_2$ | $\sigma_3$ |
|---|---|---|---|---|---|---|
| $\Gamma_{XYZ}$ | 3 | 0 | 0 | 1 | 1 | 1 |

which may be written more concisely as

$$n = \frac{1}{h} \sum N \chi_R \chi_I$$

| $C_{3v}$ | E | $2C_3$ | $3\sigma_v$ |
|---|---|---|---|
| $\Gamma_{XYZ}$ | 3 | 0 | 1 |

by the introduction of the coefficients '*N*'. As mentioned above, this simplification arises because the three reflection operations, for example, give matrices with identical characters. The representation $\Gamma_{XYZ}$ can be reduced by using the formula, or by inspection, to give the irreducible representations $A_1 + E$:

| $C_{3v}$ | E | $2C_3$ | $3\sigma_v$ | |
|---|---|---|---|---|
| $\Gamma_{XYZ}$ | 3 | 0 | 1 | |
| $A_1$ | 1 | 1 | 1 | $z$ |
| E | 2 | −1 | 0 | $(x, y)$ |

and we can identify two distinct components of this representation: $\Gamma_Z = A_1$ and $\Gamma_{XY} = E$.

We are now also in a position to understand some of the new nomenclature in the $C_{3v}$ character table. The symbol E down the left-hand column is the symbol for a *doubly degenerate representation*. This is a representation made up of $2 \times 2$ matrices, which *cannot be simplified further*. The row of numbers alongside this symbol contains the characters of these matrices, and the presence of '$(x, y)$' indicates that the functions $x$ and $y$ *together* transform as the E representation.

| $C_{3v}$ | E | $2C_3$ | $3\sigma_v$ | $h = 6$ | | |
|---|---|---|---|---|---|---|
| $A_1$ | 1 | 1 | 1 | $z$ | $x^2 + y^2, z^2$ | |
| $A_2$ | 1 | 1 | -1 | $R_z$ | | |
| E | 2 | -1 | 0 | $(x, y),$ $(R_x, R_y)$ | $(x^2 - y^2, xy),$ $(xz, yz)$ | |

Degenerate representations are encountered in point groups where there are $C_n$ or $S_n$ axes with $n = 3$ or higher. In general, for non-linear point groups, doubly degenerate representations have E as their basic symbol, to which may be added subscripts or superscripts depending on the point group. The variables $(x, y)$ are perhaps the most commonly encountered bases for E-type representations, but functions such as $(xz, yz)$ or $(x^2 - y^2, xy)$ are also important.

In linear molecules (point groups $C_{\infty v}$ or $D_{\infty h}$), doubly degenerate representations frequently have $\Pi$ (greek pi) as their principal symbol.

## 4.7 Triply degenerate representations: the point group $T_d$

Triply degenerate representations are irreducible representations composed of $3 \times 3$ matrices, and have a basic symbol T. From a chemical viewpoint, they are most frequently found in the cubic point groups $T_d$ and $O_h$, and their occurrence is best illustrated by example.

Figure 4.4 shows three vectors $x_1$, $y_1$ and $z_1$ on the central atom of a tetrahedral molecule such as $CH_4$. These vectors lie along the $C_2$ (and $S_4$) symmetry axes. The four $C_3$ axes are perpendicular to the faces of the regular tetrahedron, passing through the central atom and *one* of the hydrogen atoms (e.g. $H_1$). The planes $\sigma_d$ each contain the central carbon atom and *two* of the four hydrogen atoms. The plane shown in Fig. 4.4 bisects the angle between the $x$ and $y$ axes, and contains the atoms $H_1$ and $H_3$.

The character table for the $T_d$ point group is reproduced below:

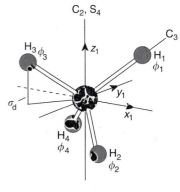

**Fig. 4.4**

| $T_d$ | E | $8C_3$ | $3C_2$ | $6S_4$ | $6\sigma_d$ | $h = 24$ | | |
|---|---|---|---|---|---|---|---|---|
| $A_1$ | 1 | 1 | 1 | 1 | 1 | | $x^2 + y^2 + z^2$ | |
| $A_2$ | 1 | 1 | 1 | -1 | -1 | | | |
| E | 2 | -1 | 2 | 0 | 0 | | $(2z^2 - x^2 - y^2, x^2 - y^2)$ | |
| $T_1$ | 3 | 0 | -1 | 1 | -1 | $(R_x, R_y, R_z)$ | | |
| $T_2$ | 3 | 0 | -1 | -1 | 1 | $(x, y, z)$ | $(xy, xz, yz)$ | |

It can be seen that there are two triply degenerate irreducible representations, $T_1$ and $T_2$, listed in the left hand column.

Alongside these symbols are the rows of numbers which correspond to characters of matrices, and in the final columns are to be found sets of functions bracketed together. The character table indicates that both

$(x, y, z)$ and $(xy, xz, yz)$ transform as $T_2$, and we shall confirm this for the $(x, y, z)$ set by working out the characters of the matrices for selected symmetry operations of the group on $x_1$, $y_1$ and $z_1$.

## 4.8    Characters for the representation $\Gamma_{xyz}$ for the central atom

We have seen previously that in order to obtain the irreducible representations for any representation $\Gamma$, it is necessary only to know the characters of the matrices which make up $\Gamma$, together with data from the character table. In this problem we will try to go directly to the characters of the various matrices, by focusing only on the diagonal terms in each $3 \times 3$ matrix, as it is only these which contribute to the character.

The identity operation leaves $x_1$, $y_1$ and $z_1$ all unshifted, and each vector therefore generates a term $+1$ on the diagonal. The character of the matrix is therefore $+3$. There are *eight* distinct $C_3$ operations to be considered in this point group: two for each of the four $C_3$ axes, but as discussed above, *any one of these* may be used to generate a matrix which will give the appropriate $\chi_R$. If we select the $C_3$ axis which lies along the $C$–$H_1$ bond (Fig. 4.4), the transformations produced by the operation $C_3^1$ are $x_1$ to $y_1$, $y_1$ to $z_1$, and $z_1$ to $x_1$. The matrix for this operation is shown alongside, and has a character zero. As indicated above, the $C_2$ and $S_4$ axes lie along the Cartesian coordinates, and if we choose the $C_2$ and $S_4$ axes lying along $z$, to be typical, we have

$C_3^1$

for $C_2(z)$: $x_1$ to $-x_1$, $y_1$ to $-y_1$, $z$ to $z_1$, leading to a character $-1$

for $S_4^1(z)$: $x_1$ to $y_1$, $y_1$ to $-x_1$, $z_1$ to $-z_1$, which also gives $-1$

Finally, as indicated previously, reflection in *any* vertical plane will result in a matrix with a character of $+1$. Here, reflection in $\sigma_d$ interchanges $x_1$ and $y_1$, but leaves $z_1$ unchanged. The matrices which form the representation $\Gamma_{xyz}$ therefore have the characters

| E | $8C_3$ | $3C_2$ | $6S_4$ | $6\sigma_d$ |
|---|---|---|---|---|
| 3 | 0 | $-1$ | $-1$ | 1 |

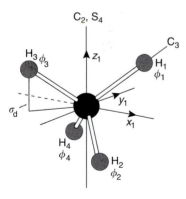

**Fig. 4.4′**

and so correspond to $T_2$.

## 4.9    The representation $\Gamma_\phi$ in $CH_4$

As a final example of the use of the reduction formula in a high symmetry point group, we shall derive the irreducible representations for the four $1s$ orbitals on the hydrogen atoms in methane. These are identified in Fig. 4.4′ as $\phi_1$ to $\phi_4$ located on the atoms $H_1$ to $H_4$. As in the example above, it is necessary only to focus on the orbitals which remain *unshifted* in order to obtain the characters of the reducible representation $\Gamma_\phi$.

The identity leaves all orbitals unshifted, resulting in a character $+4$, and a $C_3$ rotation leaves just one orbital unshifted, leading to a character $+1$. The $C_2$ and $S_4$ operations shift all the orbitals, but a typical $\sigma_d$ plane leaves two orbitals unshifted. The characters of the reducible representation $\Gamma_\phi$ are therefore

| E | $8C_3$ | $3C_2$ | $6S_4$ | $6\sigma_d$ |
|---|---|---|---|---|
| 4 | 1 | 0 | 0 | 2 |

This representation can be reduced using the formula, and shown to be $\Gamma_\phi = A_1 + T_2$, with confirmation coming from the addition of the appropriate rows in the character table.

| $T_d$ | E | $8C_3$ | $3C_2$ | $6S_4$ | $6\sigma_d$ |
|---|---|---|---|---|---|
| $A_1$ | 1 | 1 | 1 | 1 | 1 |
| $T_2$ | 3 | 0 | $-1$ | $-1$ | 1 |
| $\Gamma_\phi$ | 4 | 1 | 0 | 0 | 2 |

$$n = \frac{1}{h}\sum N_R \chi_R \chi_I$$

## 4.10  Summary

In this chapter, it has been shown that the presence of high order axes leads to *degenerate* representations, for which the smallest matrices are either $2 \times 2$ (doubly degenerate) or $3 \times 3$ (triply degenerate). Examples of the occurrence of these representations are presented and discussed.

## 4.11  Exercises

1.  The molecules $NH_3$, $BF_3$ and $ClF_3$ have symmetries $C_{3v}$, $D_{3h}$ and $C_{2v}$ (T-shaped) respectively. Derive the irreducible representations $\Gamma_{N-H}$, $\Gamma_{B-F}$ and $\Gamma_{Cl-F}$ in these molecules.
2.  Derive the irreducible representations $\Gamma_{Xe-O}$ and $\Gamma_{Xe-F}$ in the molecule $XeOF_4$ (see Fig. 1.24).
3.  Derive the irreducible representations $\Gamma_{B-Cl}$ and $\Gamma_{B-B}$ in $B_2Cl_4$ (see Fig. 1.16).
4.  Show that the irreducible representation $\Gamma_{S-F}$ in the molecule $SF_6$ is $\Gamma_{S-F} = A_{1g} + E_g + T_{1u}$.

# 5 Molecular vibrations
# (non-degenerate modes)

One of the fundamental aims of this Primer was to suggest that symmetry, and in particular the symmetry of molecular species, is a sufficiently interesting subject to merit further study. We saw that molecular shapes could be described and classified in terms of the molecular point group, and that the inter-relation between symmetry operations led to the idea of a representation, which could be illustrated by atomic and molecular features such as orbitals, bonds or displacement vectors.

In this chapter, the main aim will be to understand the role of symmetry in describing molecular vibrations, and a suitable starting point is to consider the '$3n$' degrees of freedom of motion in a molecule in relation to any symmetry elements which may be present.

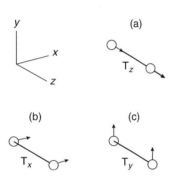

Fig. 5.1

## 5.1 Description of atomic and molecular displacements

An isolated atom has three degrees of freedom. These may be represented by translational displacements in three mutually perpendicular directions.

When two atoms combine to form a diatomic molecule, the total number of degrees of freedom is now six, but three types of displacement can be identified: translations, rotations and a vibration. The 'in-phase' displacements of the atoms along the Cartesian directions result in net translations of the whole molecule, with no change in internuclear distance or in molecular orientation. Typical 'in-phase' displacements along $z$, $x$ and $y$ are shown in Figs 5.1(a)–(c). They represent the three 'translational degrees of freedom'. Counterparts of this in-phase motion are to be found in all molecules, and these combinations of displacements can be labelled as $T_x$, $T_y$ or $T_z$, where $T_x$ indicates overall translation in the $x$ direction.

Figure 5.2(a) shows the result of 'out-of-phase' motion along the $x$ direction. Here, the centre of mass of the molecule is unshifted, and there is no change in internuclear distance. However, there has been a change in orientation with respect to the external frame of reference, and this combined motion of the two atoms corresponds to rotation about the $y$ axis. This can be described as the rotational degree of freedom $R_y$. In a similar way, the atomic displacements in Fig. 5.2(b) correspond to $R_x$. Rotation about the internuclear axis in a linear molecule such as shown here does not constitute a degree of freedom, but all *non-linear* molecules have a total of three rotational degrees of freedom, $R_x$, $R_y$ and $R_z$.

The third example of 'out-phase' motion takes place along the $z$ axis, and is shown in Fig. 5.2(c). This motion is clearly a vibration, in which the

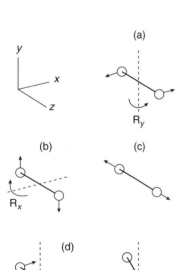

Fig. 5.2

centre of mass and overall orientation remain unchanged, but there has been a change in an *internal coordinate* of the molecule—in this case the internuclear distance.

Finally, we should consider briefly the effect of independent motion of the two atoms in our molecule, such as might be visualised by Fig. 5.2(d), where the two atoms move along different axes. This motion results in a shift in the centre of mass, a change in internuclear distance, and also a change in orientation. However, despite this apparent complexity, the overall effect can be shown to be made up of a sum of components of motion already described.

In general, all linear molecules have three translational degrees of freedom and two rotational degrees of freedom, with the result that they have a total of $3n - 5$ vibrations. Non-linear molecules have three translation and three rotational degrees of freedom, and hence $3n - 6$ vibrations, and it is the *symmetries* of these translations, rotations and vibrations that we now need to establish.

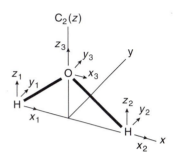

**Fig. 5.3**

## 5.2 Atomic displacement coordinates as bases for the representation of molecular motion: the H₂O molecule

In order to determine the symmetries of the various translational, rotational and vibrational degrees of freedom in a molecule, we need to work out the irreducible representations which arise from the use of $x$, $y$ and $z$ displacement coordinates on all the atoms present.

For a molecule containing $n$ atoms, this will involve $3n$ such coordinates, and Fig. 5.3 provides a suitable framework for this procedure in the case of the H₂O molecule.

However, before tackling this problem from first principles, it is helpful to look at ways in which the complete representation, which we can denote as $\Gamma_{mol}$, can be built up by considering the contributions made to the representation firstly by the two hydrogen atoms, and then by the oxygen atom.

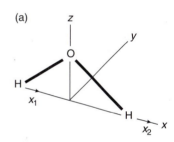

## 5.3 Representations for H atom and O atom motion

Figure 5.4(a) shows the vectors $x_1$ and $x_2$ on the two hydrogen atoms. As we have seen earlier, the H₂O molecule has C₂ᵥ symmetry, and in order to obtain the representations using these vectors as bases, we have to consider the characters of the matrices which describe the effect of the various symmetry operations (E, C₂, etc.) on this pair of vectors, remembering that only vectors which remain *unshifted* will contribute to the character.

The operations E and $\sigma_v(xz)$ both leave these vectors unshifted, and so result in matrices with characters of +2, whilst for the operations C₂ and $\sigma_v'(yz)$ the resulting characters are zero.

The reducible representation therefore looks like:

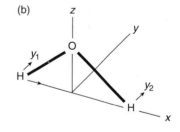

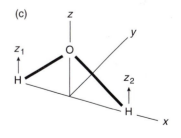

| C₂ᵥ | E | C₂ | $\sigma_v(xz)$ | $\sigma_v'(yz)$ |
|---|---|---|---|---|
| $\Gamma_{H(x)}$ | 2 | 0 | 2 | 0 |

**Fig. 5.4**

Reference to the $C_{2v}$ character table shows that this can be reduced to $A_1+B_1$.

In a similar way, Figs 5.4(b) and (c) show the $y$ and $z$ vectors respectively for the two hydrogen atoms, and the representations which arise are

| $C_{2v}$ | E | $C_2$ | $\sigma_v(xz)$ | $\sigma_v'(yz)$ | | |
|---|---|---|---|---|---|---|
| $\Gamma_{H(y)}$ | 2 | 0 | $-2$ | 0 | $=$ | $A_2 + B_2$ |

and

| | | | | | | |
|---|---|---|---|---|---|---|
| $\Gamma_{H(z)}$ | 2 | 0 | 2 | 0 | $=$ | $A_1 + B_1$ |

The irreducible representations for hydrogen atom motion are therefore: $\Gamma_{H(xyz)} = 2A_1 + A_2 + 2B_1 + B_2$.

Figure 5.5 shows the $x$, $y$ and $z$ vectors on the oxygen atom. This atom lies on all the symmetry elements, but some operations result in a reversal of vector direction (giving a contribution of $-1$ to the character) whilst others leave both the position and direction unchanged. The representation for oxygen atom motion turns out to be

| $C_{2v}$ | E | $C_2$ | $\sigma_v(xz)$ | $\sigma_v'(yz)$ | | |
|---|---|---|---|---|---|---|
| $\Gamma_{O(xyz)}$ | 3 | $-1$ | 1 | 1 | $=$ | $A_1 + B_1 + B_2$ |

a result which is identical to that previously derived for an atom at the coordinate origin (Chapter 3, Fig. 3.3).

The total representation for atomic motion in the $H_2O$ molecule is obtained by building up the contributions *from the two sets of atoms*, and is therefore

$$\Gamma_{mol} = \Gamma_H + \Gamma_O = 3A_1 + A_2 + 3B_1 + 2B_2$$

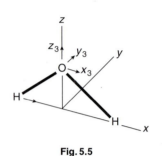

**Fig. 5.5**

### 5.4  $\Gamma_{mol}$ from first principles for $H_2O$—the significance of unshifted atoms

The derivation of $\Gamma_{mol}$ directly by considering the effect of the symmetry operations on all the atom vectors simultaneously is at first sight a daunting prospect—even for a triatomic molecule such as $H_2O$. What is required is the derivation of four $9 \times 9$ matrices which describe the effect of the four symmetry operations in the $C_{2v}$ point group on the nine vectors $x_1$, $y_1$, $z_1$, $x_2$, $y_2$, etc. shown in Fig. 5.6.

The simplest of these matrices arises from the effect of the identity operation. This operation leaves all the vectors unchanged, and results in a $9 \times 9$ matrix with a character of 9 (Fig. 5.7).

In contrast, the $C_2$ operation (Fig. 5.8) leaves only one vector unchanged ($z_3$), reverses $x_3$ and $y_3$, but *shifts all the vectors on the H atoms*, e.g. $C_2(z_1) \rightarrow (z_2)$.

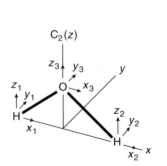

**Fig. 5.6**

$$E \begin{pmatrix} x_1 \\ y_1 \\ z_1 \\ x_2 \\ y_2 \\ z_2 \\ x_3 \\ y_3 \\ z_3 \end{pmatrix} = \begin{pmatrix} 1 & 0 & 0 & 0 & 0 & 0 & 0 & 0 & 0 \\ 0 & 1 & 0 & 0 & 0 & 0 & 0 & 0 & 0 \\ 0 & 0 & 1 & 0 & 0 & 0 & 0 & 0 & 0 \\ 0 & 0 & 0 & 1 & 0 & 0 & 0 & 0 & 0 \\ 0 & 0 & 0 & 0 & 1 & 0 & 0 & 0 & 0 \\ 0 & 0 & 0 & 0 & 0 & 1 & 0 & 0 & 0 \\ 0 & 0 & 0 & 0 & 0 & 0 & 1 & 0 & 0 \\ 0 & 0 & 0 & 0 & 0 & 0 & 0 & 1 & 0 \\ 0 & 0 & 0 & 0 & 0 & 0 & 0 & 0 & 1 \end{pmatrix} \begin{pmatrix} x_1 \\ y_1 \\ z_1 \\ x_2 \\ y_2 \\ z_2 \\ x_3 \\ y_3 \\ z_3 \end{pmatrix}$$

**Fig. 5.7**

$$C_2 \begin{pmatrix} x_1 \\ y_1 \\ z_1 \\ x_2 \\ y_2 \\ z_2 \\ x_3 \\ y_3 \\ z_3 \end{pmatrix} = \begin{pmatrix} 0 & 0 & 0 & -1 & 0 & 0 & 0 & 0 & 0 \\ 0 & 0 & 0 & 0 & -1 & 0 & 0 & 0 & 0 \\ 0 & 0 & 0 & 0 & 0 & 1 & 0 & 0 & 0 \\ -1 & 0 & 0 & 0 & 0 & 0 & 0 & 0 & 0 \\ 0 & -1 & 0 & 0 & 0 & 0 & 0 & 0 & 0 \\ 0 & 0 & 1 & 0 & 0 & 0 & 0 & 0 & 0 \\ 0 & 0 & 0 & 0 & 0 & 0 & -1 & 0 & 0 \\ 0 & 0 & 0 & 0 & 0 & 0 & 0 & -1 & 0 \\ 0 & 0 & 0 & 0 & 0 & 0 & 0 & 0 & 1 \end{pmatrix} \begin{pmatrix} x_1 \\ y_1 \\ z_1 \\ x_2 \\ y_2 \\ z_2 \\ x_3 \\ y_3 \\ z_3 \end{pmatrix}$$

**Fig. 5.8**

This has the effect of placing a *zero* on the diagonal for this and all other operations which similarly *move* a vector, and it arises because the hydrogen atoms themselves interchange position as a result of the $C_2$ operation.

Only the three vectors $x_3$, $y_3$ and $z_3$ therefore contribute to the character, giving a net effect of $-1$.

The position regarding reflection in $\sigma_v(yz)$ is similar. Here, the H atoms again interchange position, with the result that their vectors do not contribute to the character. Only the O atom remains unshifted, and the contributions to the character from $x_3$, $y_3$ and $z_3$ are now $-1$, $+1$ and $+1$, giving a character of $+1$.

Reflection in the plane of the molecule, $\sigma_v(xz)$, leaves all the atoms unshifted, but reverses the $y$ vectors whilst leaving $x$ and $z$ unchanged. Each unshifted atom therefore contributes $+1$ to the character, and as this operation in fact leaves three atoms unshifted, the total contribution to the character is $+3$.

The characters of the reducible representation derived in this way are therefore seen to be

| $C_{2v}$ | E | $C_2$ | $\sigma_v(xz)$ | $\sigma_v'(yz)$ |
|---|---|---|---|---|
| $\Gamma_{mol}$ | 9 | $-1$ | 3 | 1 |

This can be reduced using the reduction formula to give

$$\Gamma_{mol} = 3A_1 + A_2 + 3B_1 + 2B_2$$

a result identical to that above.

The main conclusions which therefore emerge are:

$$n = \frac{1}{h} \sum N \chi_R \chi_I$$

1. Only the vectors on unshifted atoms can contribute to the character of the reducible representation.
2. The size of the contribution depends on the type of symmetry element (e.g. E, $C_2$, etc).

### 5.5   Translational, rotational and vibrational symmetry in $H_2O$

The result we have obtained above classifies the nine degrees of freedom in the $H_2O$ molecule in terms of their symmetries, and the first step in identifying the symmetries of the vibrations is to recognise that the total representation, $\Gamma_{mol}$, is made up of the separate representations for translation, rotation and vibration. This may be written as

$$\Gamma_{mol} = \Gamma_{trans} + \Gamma_{rot} + \Gamma_{vib}$$

and if our primary aim is to discover the symmetries of the molecular vibrations, it is evident that

$$\Gamma_{vib} = \Gamma_{mol} - \Gamma_{trans} - \Gamma_{rot}$$

From above, $\Gamma_{mol} = 3A_1+A_2+3B_1+2B_2$, and all that is required in order to discover which of these representations describe the symmetries of the vibrations are the irreducible representations which make up $\Gamma_{trans}$ and $\Gamma_{rot}$. This information is readily available from the $C_{2v}$ character table.

| $C_{2v}$ | E | $C_2$ | $\sigma_v(xz)$ | $\sigma_v'(yz)$ | $h = 4$ | |
|---|---|---|---|---|---|---|
| $A_1$ | 1 | 1 | 1 | 1 | $z$ | $x^2, y^2, z^2$ |
| $A_2$ | 1 | 1 | $-1$ | $-1$ | $R_z$ | $xy$ |
| $B_1$ | 1 | $-1$ | 1 | $-1$ | $x, R_y$ | $xz$ |
| $B_2$ | 1 | $-1$ | $-1$ | 1 | $y, R_x$ | $yz$ |

The $H_2O$ molecule is non-linear, and so the representation $\Gamma_{rot}$ has three contributions, $R_x$, $R_y$ and $R_z$, corresponding to rotation about each of the principal axes. The symmetries of $R_x$, $R_y$ and $R_z$ are given in the penultimate column of the $C_{2v}$ character table, where it can be seen that they belong to the representations $B_2$, $B_1$ and $A_2$ respectively. The rotational degrees of freedom are

$$\Gamma_{rot} = A_2 + B_1 + B_2$$

The representations which contribute to $\Gamma_{trans}$ may also be obtained from the character table. In this case, it is the functions $x$, $y$ and $z$ which point to the representations for translational motion. These functions also appear in the penultimate column of the character table. From the $C_{2v}$ character table, we therefore have

$$\Gamma_{trans} = A_1 + B_1 + B_2$$

In some editions of character tables, the functions $x$, $y$, and $z$ are listed as $T_x$, $T_y$, and $T_z$, which reinforces this description.

We are now in a position to derive the symmetries of the vibrational modes in the $H_2O$ molecule:

$$\Gamma_{vib} = \Gamma_{mol} - \Gamma_{trans} - \Gamma_{rot}$$
$$= (3A_1 + A_2 + 3B_1 + 2B_2) - (A_1 + B_1 + B_2) - (A_2 + B_1 + B_2)$$

i.e.   $\Gamma_{vib} = 2A_1 + B_1$

## 5.6 Pictorial representation of translations, rotations and vibrations in the $H_2O$ molecule

The expressions we have obtained above for the symmetries of the various degrees of freedom in $H_2O$ initially give little indication of the types of atom motion involved, and it is often useful to return to the original cartesian vector displacements to visualise what is happening. This will also illustrate the essential features of $A_1$- and $B_1$-type vibrations. The translational degrees of freedom are represented by motion of the whole molecule along the three cartesian axes, as shown in Fig. 5.9, and the rotations $R_x$ and $R_z$ are similarly easy to visualise (Fig. 5.10). However, the remaining rotational degree of freedom, $R_y$, and the three vibrations require more investigation.

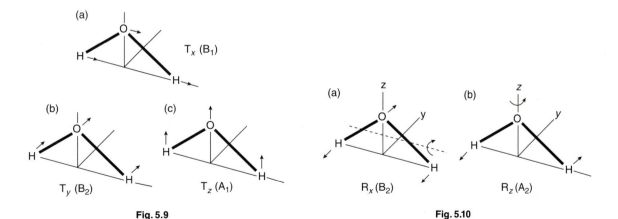

Fig. 5.9   Fig. 5.10

The symmetries of these representations are $B_1$ ($R_y$) and $2A_1+B_1$ (vibrations), and are such that atom motion in these degrees of freedom takes place in the ($xz$) plane of the molecule. Figures 5.11(a) and (b) show two sets of vector displacements which have $B_1$ symmetry, and which correspond *closely, but not exactly*, to rotational and vibrational motion respectively.

If we look firstly at Fig. 5.11(a), it is clear that although the motion is predominantly a rotation about the $y$ axis, displacements in the directions shown will alter the lengths of the bonds, and also shift the centre of mass of the molecule. It is not therefore a pure rotation. In the same way, the atom displacements in Fig. 5.11(b) result mainly in changes in bond length, but there is also some rotational motion.

Neither of these pictures corresponds precisely to a vibration or a rotation, and the actual displacements which take place in these two degrees of freedom are more likely to resemble the vectors shown in the related diagrams 5.11(c) and (d). However, all four pictures in this figure exhibit an essential feature of $B_1$-type symmetry, which is a basic asymmetry along the $x$ axis.

Figures 5.12(a) and (b) show the idealised motion for the two $A_1$ vibrations which can be constructed from the available Cartesian vectors. Vibrations such as these, which belong to the totally symmetric representation, are often described as 'totally symmetric modes', and their distinguishing feature is that, at all stages in the vibration, the molecule retains all the symmetry elements present in the equilibrium structure.

However, as we found for the $B_1$ displacements above, at least one of the depicted motions (Fig. 5.12(a)) cannot be the pure vibration expected, as it also involves a change in the centre of mass. Again, the accompanying diagrams (c) and (d) suggest more realistic pictures of actual atom displacements.

Fortunately, the use which we shall make of the symmetry of a vibration *does not depend* on knowing the precise displacements of the atoms, and, indeed, these are only rarely calculated. All that we shall need is the symmetry.

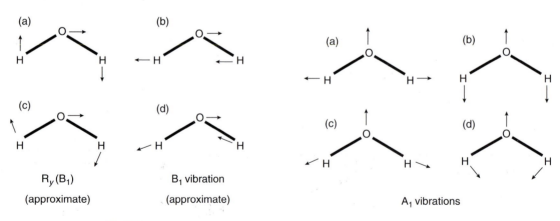

(a)

(c)

$R_y(B_1)$

(approximate)

(b)

(d)

$B_1$ vibration

(approximate)

**Fig. 5.11**

(a)

(c)

(b)

(d)

$A_1$ vibrations

**Fig. 5.12**

## 5.7 Contributions to the character from unshifted atoms— other symmetry operations

An important step in the determination of $\Gamma_{mol}$ for the $H_2O$ molecule was the fact that it was necessary to consider only those atoms which were *unshifted* by a symmetry operation when working out the characters of the associated matrices. In particular, we saw that the contributions *per unshifted atom* to the matrices for the operations E, $C_2$, $\sigma_v(xz)$ and $\sigma_v'(yz)$ are +3, −1, +1 and +1 respectively.

The stage has now been reached when we need to consider the contributions from other symmetry operations so that we can tackle almost any problem.

In Chapter 4, we saw that the effect of a $C_n$ axis on a point $(X, Y, Z)$ results in a contribution to the character of the matrix describing the operation $C_n^1$ which is given by

$$\chi C_n^1 = 2\cos\theta + 1$$

*This equation also defines the contribution made by a simple rotation to the character of $\Gamma_{mol}$ for each unshifted atom.*

Thus, for the $C_2$ operation, $\theta = 180°$, $\cos(180°) = -1$, and the total contribution to the character is −1 per unshifted atom. For $C_4^1$, $\cos\theta = 0$, and the contribution is +1.

In this equation, the term '$2\cos\theta$' arises from rotation in the $xy$ plane, and the term '+1' takes account of the fact that with the $C_n$ axis coincident with $z$, the operation $C_n^1$ leaves any vector along $z$ unshifted.

If instead of having the operation $C_n^1$, we consider the rotation–reflection operation $S_n^1$, we can visualise that the contribution in the $xy$ plane will be the same, but that the 'reflection' component of this operation will now *reverse* any vector along $z$. The corresponding equation for an $S_n^1$ operation will therefore be

$$\chi S_n^1 = 2\cos\theta - 1$$

We are now in a position to derive the contributions per unshifted atom to $\Gamma_{mol}$ for all $C_n$ and $S_n$ symmetry operations, and examples of these are summarised in Table 5.1. Also included in this table, for completeness, are the contributions arising from the identity (E), from an atom lying at a centre of inversion (i), and from an atom on any plane ($\sigma$).

**Table 5.1** Contributions to characters of unshifted atoms for typical symmetry operations

| Symmetry operation | Contribution to character |
|---|---|
| E (= $C_1$) | 3 |
| i (= $S_2$) | −3 |
| $\sigma$ (= $S_1$) | 1 |
| $C_2$ | −1 |
| $C_3$ | 0 |
| $C_4$ | 1 |
| $C_6$ | 2 |
| $S_3$ | −2 |
| $S_4$ | −1 |
| $S_6$ | 0 |

## 5.8 General procedure for determining the symmetries of molecular vibrations

Using the relationship

$$\Gamma_{vib} = \Gamma_{mol} - \Gamma_{trans} - \Gamma_{rot}$$

we can now identify all the steps necessary in the determination of the symmetries of the vibrational modes in a molecule.

1. Identify the positions of the symmetry elements and establish the molecular point group.
2. Work out the characters of the matrices which make up $\Gamma_{mol}$. This problem is most conveniently solved using the 'unshifted atom' approach, taking account of the contributions made to the character by those symmetry operations which leave atoms unshifted. Table 5.1 summarises the contribution per unshifted atom for the most commonly encountered symmetry operations.
3. Use the reduction formula to obtain the irreducible representations present in $\Gamma_{mol}$.
4. Use the character table to identify the irreducible representations for translation and rotation—i.e. $\Gamma_{trans}$ and $\Gamma_{rot}$—and subtract these representations from $\Gamma_{mol}$.

$$n = \frac{1}{h}\sum N\chi_R\chi_I$$

The representations remaining correspond to $\Gamma_{vib}$.

### 5.9   Illustration: to determine the symmetries of the vibrational modes in molecular $SO_2F_2$

*Step 1*: The molecule $SO_2F_2$ has $C_{2v}$ symmetry, with a shape based on tetrahedral coordination round the central sulphur atom. Figure 5.13 shows the basic structure in a coordinate framework which places the sulphur atom at the origin, one pair of equivalent atoms (e.g the F atoms) in the plane $\sigma_v(xz)$, and the other pair in the $\sigma'_v(yz)$ plane. Only the sulphur atom lies on the $C_2$ axis.

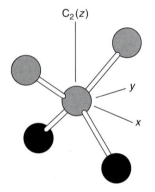
$C_2(z)$

**Fig. 5.13**

| $C_{2v}$ | E | $C_2$ | $\sigma_v(xz)$ | $\sigma'_v(yz)$ | $h = 4$ | |
|---|---|---|---|---|---|---|
| $A_1$ | 1 | 1 | 1 | 1 | $z$ | $x^2, y^2, z^2$ |
| $A_2$ | 1 | 1 | −1 | −1 | $R_z$ | $xy$ |
| $B_1$ | 1 | −1 | 1 | −1 | $x, R_y$ | $xz$ |
| $B_2$ | 1 | −1 | −1 | 1 | $y, R_x$ | $yz$ |

*Step 2*: In order to work out $\Gamma_{mol}$, it is then often convenient to construct a small table which identifies firstly the number of atoms which are *unshifted* by each of the symmetry operations, and secondly the contribution made to the character *per unshifted atom*. The contributions to $\Gamma_{mol}$ for each symmetry operation are then obtained by multiplication. For $SO_2F_2$, such a table would take the form:

| Symmetry operation | E | $C_2$ | $\sigma_v(xz)$ | $\sigma'_v(yz)$ |
|---|---|---|---|---|
| Number of unshifted atoms | 5 | 1 | 3 | 3 |
| Contribution per unshifted atom | 3 | −1 | 1 | 1 |
| Character in $\Gamma_{mol}$ | 15 | −1 | 3 | 3 |

*Step 3*: Application of the reduction formula then gives

$$n = \frac{1}{h}\sum N_R \chi_R \chi_I$$

$$n(A_1) = (1/4)[(15 \times 1 \times 1) + (-1 \times 1 \times 1) + (3 \times 1 \times 1) + (3 \times 1 \times 1)]$$
$$= 5$$

$$n(A_2) = (1/4)[(15 \times 1 \times 1) + (-1 \times 1 \times 1) + (3 \times -1 \times 1) + (3 \times -1 \times 1)]$$
$$= 2$$

$$n(B_1) = (1/4)[(15) + (-1 \times -1 \times 1) + (3 \times 1 \times 1) + (3 \times -1 \times 1)] = 4$$

$$n(B_2) = (1/4)[(15) + (-1 \times -1 \times 1) + (3 \times -1 \times 1) + (3 \times 1 \times 1)] = 4$$

and hence

$$\Gamma_{mol} = 5A_1 + 2A_2 + 4B_1 + 4B_2$$

*Step 4*: From the $C_{2v}$ character table, the translations and rotations have symmetries

$$\Gamma_{trans} = A_1 + B_1 + B_2 \quad \text{and} \quad \Gamma_{rot} = A_2 + B_1 + B_2$$

Since $\Gamma_{vib} = \Gamma_{mol} - \Gamma_{trans} - \Gamma_{rot}$, this leads to the final result:

$$\Gamma_{vib} = 4A_1 + A_2 + 2B_1 + 2B_2$$

## 5.10 Molecular vibrations and internal coordinates

The procedure described above for deriving the symmetries of molecular vibrations is based on an approach which initially considers all possible degrees of freedom in a molecule using a set of $3n$ 'external' Cartesian coordinates, and vibrations emerge as the type of motion remaining when molecular translations and rotations have been accounted for.

This approach leads directly to the symmetries of all the vibrational modes in a molecule, but is not convenient for describing such vibrations in terms of simple concepts such as changes in bond lengths or bond angles.

In order to express molecular vibrations in this more pictorial way, we need instead to use 'internal coordinates'—bond lengths and angles—and in particular to use these coordinates as bases from which to derive representations for 'stretching vibrations' or 'bending vibrations'.

The justification for this distinction between 'stretches' and 'bends' is largely pragmatic, and is well-illustrated by the 'ball and spring' view of molecular structure. In general, more energy is required to stretch a bond than to change an interbond angle, and unless there is a wide disparity in atomic mass, stretching frequencies generally occur at higher energies than bends.

### Internal coordinates in $H_2O$, $NH_3$ and $CH_4$

For simple molecules such as $H_2O$ or $NH_3$, it turns out that the $3n-6$ vibrations can readily be described in terms of clearly identifiable bond

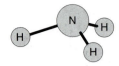

Fig. 5.14

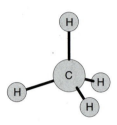

Fig. 5.15

lengths and angles. In the case of $H_2O$, there are three vibrations, and these can be visualised as changes in the three important internal coordinates— the two O–H bond lengths and the H–O–H bond angle. In $NH_3$, $n = 4$; there are six vibrational degrees of freedom, and it can be shown that the vibrations can similarly be described in terms of changes to the three N–H bonds and three H–N–H angles (Fig. 5.14).

For $CH_4$, however, the situation is more complex. The molecule has *nine* vibrational degrees of freedom $(3n - 6 = 9)$, but there are available apparently *ten* valid internal coordinates—made up of the four C–H bonds and the six H–C–H angles (see Fig. 5.15). The simple correspondence between the number of vibrational modes and the number of internal coordinates therefore no longer applies. The explanation, and method of dealing with this anomaly are beyond the scope of this Primer, but the outcome is that we can retain the *four C–H bonds* as bonafide internal coordinates to describe the *stretching vibrations* in $CH_4$.

*Suitability of internal coordinates to describe vibrations*: From what has been suggested above, the use of bond angles as internal coordinates for bending modes may cause problems with some molecular shapes. However in general, *the use of bond lengths as bases for stretching modes is acceptable*.

In the final part of this chapter, the approach adopted will therefore be to derive the symmetries of molecular stretching vibrations using bonds as bases for representations, and to obtain the symmetries of the bending modes by defining all non-stretching modes as 'bends', and using the relationship

$$\Gamma_{vib} = \Gamma_{stretch} + \Gamma_{bend}$$

This relationship will not distinguish between the different types of bending motion which may be present in a molecule (e.g. 'out-of-plane' or 'in-plane' bending in a planar molecule such as $C_2H_4$), but will be sufficient at this stage.

### 5.11   The symmetries of stretching vibrations

As indicated above, the symmetries of the stretching vibrations $\Gamma_{stretch}$ in a molecule can be obtained from the representation derived from bonds. For molecules which contain more than one kind of bond, such as $SO_2F_2$, the stretching modes can be subdivided into two types: the S–O stretching modes and the S–F stretches.

In order to demonstrate the general procedure for obtaining the symmetries of stretching modes, it is convenient first to look again at two molecules $H_2O$ and $SO_2F_2$ for which the complete vibrational representation $\Gamma_{vib}$ has already been obtained.

### Stretching modes in $H_2O$

The symmetries of the stretching modes in the $H_2O$ molecule are obtained by reducing the representation $\Gamma$ which relates to the two O–H bonds. The

$H_2O$ molecule has $C_{2v}$ symmetry, and if we identify the bonds as $r_1$ and $r_2$ (Fig. 5.16), the characters of the reducible representation $\Gamma_r$ are

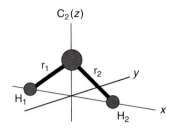

Fig. 5.16

$$
\begin{array}{cccc}
E & C_2 & \sigma_v(xz) & \sigma_v'(yz) \\
\Gamma_r \quad 2 & 0 & 2 & 0
\end{array}
$$

This may then be reduced to

$$\Gamma_r = \Gamma_{stretch} = A_1 + B_1$$

This result, which was derived in Chapter 3 as a simple exercise, now takes on more significance. These irreducible representations describe the symmetries of the two stretching modes in $H_2O$.

We have seen previously that the $H_2O$ molecule has a total of three vibrations $2A_1 + B_1$, and the approximate atomic motions in these vibrations are reproduced in Fig. 5.17. This independent derivation of the symmetries of the stretching modes allows us to use the relationship

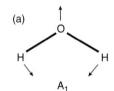

$$\Gamma_{vib} = \Gamma_{stretch} + \Gamma_{bend} = 2A_1 + B_1$$

from which $\Gamma_{bend} = A_1$.

From experiment, the $H_2O$ vibrational modes are found at 3756, 3657 and 1595 cm$^{-1}$. One mode is thus at a significantly lower frequency than the other two, and is identified as the $A_1$ bending mode (Fig. 5.17(a)). The two high frequency modes correspond to the two stretches, and it can be shown that the 3657 cm$^{-1}$ vibration is the one which has $A_1$ symmetry, involving the simultaneous stretching of both bonds (Fig. 5.17(b)), whilst the 3756 cm$^{-1}$ mode has symmetry $B_1$. Here, one bond lengthens whilst the other shortens (Fig. 5.17(c)), and this type of motion is often described as an 'asymmetric stretch'.

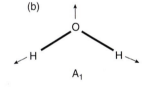

These results for the $H_2O$ molecule also illustrate two important general results: Firstly, although the two O–H bonds in the water molecule are equivalent, the symmetric ($A_1$) and antisymmetric ($B_1$) stretches are found to have *different* vibration frequencies, and in general it is found that *each vibrational mode in a molecule has a unique frequency*.

Secondly, the use of the *two* bonds $r_1$ and $r_2$ as bases for the reducible representation of stretching modes has produced *two* irreducible representations, in this case $A_1 + B_1$. This correspondence between the number of bonds and the number of related irreducible representations is also a quite general result. In using this relationship, however, account has to be taken of any degeneracies which might be present (as we shall see below), and the term 'bond' should be interpreted simply as an internuclear distance, with no implication of bond order. Here, the C–C bond in $C_2H_4$ would be regarded as a single internuclear distance, not as separate $\sigma$ and $\pi$ bonds! With these provisos, *the number of stretching modes in a molecule is equal to the number of bonds*.

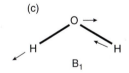

Fig. 5.17 Vibrations of $H_2O$

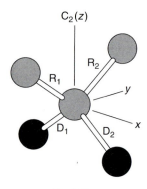

Fig. 5.18

## Stretching modes in SO₂F₂

The molecule $SO_2F_2$ has $C_{2v}$ symmetry, and the representations for all the vibrations have been derived as being

$$\Gamma_{vib} = 4A_1 + A_2 + 2B_1 + 2B_2$$

There are two S–O bonds and two S–F bonds, and these may be identified as $R_1$, $R_2$, $D_1$ and $D_2$ as indicated in Fig. 5.18.

These four bonds will give rise to four stretching modes, but because there are two distinct types of bond, R and D, the stretching modes can be subdivided into two sets, $\Gamma_R$ and $\Gamma_D$.

The characters of the reducible representations $\Gamma_R$ and $\Gamma_D$ are obtained in the usual way by noting how many of the bonds within a set remain *unshifted* by the symmetry operations in $C_{2v}$:

|            | E | $C_2$ | $\sigma_v(xz)$ | $\sigma_v'(yz)$ |
|------------|---|-------|----------------|-----------------|
| $\Gamma_R$ | 2 | 0     | 0              | 2               |
| $\Gamma_D$ | 2 | 0     | 2              | 0               |

and the corresponding irreducible representations are then obtained either by inspection, or by using the reduction formula

$$\Gamma_R = A_1 + B_2, \qquad \Gamma_D = A_1 + B_1$$

Within the total representation $\Gamma_{vib}$, the four stretching modes contribute $2A_1+B_1+B_2$, and the representations remaining correspond to bending modes:

$$\Gamma_{bend} = 2A_1 + A_2 + B_1 + B_2$$

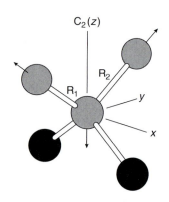

Fig. 5.19(a) Symmetric S–O stretch.

From experiment, the four stretching modes have been identified at 1502, 1269, 885 and 848 $cm^{-1}$. The higher frequency pair corresponds to two S–O stretches and the other pair to two S–F stretches. Within each of these pairs, there is a 'symmetric stretch' and an 'asymmetric stretch', and the atom displacements associated with the S–O stretching modes are indicated in Figs. 5.19(a) and (b). In practice, as we saw previously for the $H_2O$ molecule, these vectors give only an approximate picture of true atom displacements, but are sufficient to illustrate the symmetries of the different modes.

*A third important general result*: One result which has emerged from deriving the symmetries of the stretching modes in $H_2O$ and in $SO_2F_2$ is that each set of equivalent bonds, r, R or D, has produced a totally symmetric representation—in these examples an $A_1$, since both molecules belong to the $C_{2v}$ point group. This is an example of a quite general result: *it is always possible to stretch a set of equivalent bonds such that the full symmetry of the molecule is preserved.*

Because of this, *the representation $\Gamma_r$ obtained from a set of equivalent bonds (r) will always contain the totally symmetric representation.*

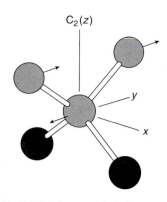

Fig. 5.19(b) Asymmetric S–O stretch.

## 5.12 Recipe for deriving the symmetries of stretching modes

We are now in a position to summarise the essential steps in deriving the symmetries of the stretching vibrations for any molecule.

*Step 1*: Identify the symmetry elements present in the molecule, and derive the point group.

*Step 2*: Consult the appropriate character table to check (a) that the orientation of the molecule is consistent with the axis system used in the character table, and (b) that all the symmetry operations listed can be identified in the molecule.

*Step 3*: Identify and label the sets of equivalent bonds.

*Step 4*: Work out the characters of the reducible representation for each set of equivalent bonds, by considering those bonds which remain unshifted by a particular symmetry operation.

$$n = \frac{1}{h} \sum N_\chi R \chi_I$$

*Step 5*: Obtain the irreducible representations for each set of bonds, either by using the reduction formula or by inspection. Note that for each set of bonds there will be one totally symmetric representation.

The resulting representations describe the symmetries of the stretching modes.

## 5.13 Summary

The material in this chapter has hopefully provided sufficient detail for a basic treatment of molecular vibrations in terms of their symmetry properties.

As indicated earlier, the *actual atomic displacements* which take place during a vibration are *rarely calculated* on a routine basis. These depend not only on the symmetry of the vibration, but upon relative atomic masses, bond angles, electron distribution within the molecule, and, in many molecules, there is also a 'mixing' of vibrational modes of the same symmetry.

The subject matter in this chapter leads directly into Chapter 6, and to the relationship between symmetry and vibrational spectroscopy. The exercises at the end of the next chapter relate to material in both Chapters 5 and 6.

# 6 Vibrational spectroscopy— degenerate vibrations

One of the main purposes behind identifying the symmetries of different molecular vibrations lies in the relationship between vibrational symmetry and spectroscopic methods of characterisation. This link between molecular symmetry and spectroscopic observations relies on the fact that the number and type of vibrational energy levels which exist for a molecule depend upon its symmetry, and that the spectroscopic 'selection rules', which dictate which transitions between these levels we can expect to observe, also depend upon molecular symmetry.

The link has been described in detail elsewhere (see Appendix III), but it is useful at this stage to state the basic *symmetry selection rules* which determine whether a particular vibration can be detected using the two commonly encountered techniques of infrared (IR) and Raman spectroscopy.

## 6.1  IR spectroscopy

Molecular infrared spectroscopy, in its most frequently encountered form, involves the *absorption* of electromagnetic radiation at an energy which corresponds to a vibrational transition within the molecule. However, in general, not all molecular vibrations give rise to such an absorption of energy. Only those vibrations with *particular symmetries* can absorb infrared radiation, and hence be detected using the technique.

The circumstances which define which molecular vibrations can be detected using IR spectroscopy—i.e those vibrations which are *infrared active*—are expressed as a *symmetry selection rule* for IR spectroscopy, which states that

*A vibration will be active in the infrared if it belongs to the same representation as one of the functions x, y or z.*

In a slightly less precise form, this selection rule is often encountered as

*A vibration will be active in the infrared if it has the same symmetry as one of the functions x, y or z*

At a more descriptive level it is

*A vibration will be active in the infrared if it involves a change in dipole moment of the molecule.*

A major part of this Primer has been to introduce and illustrate the concept of a representation, and we have also learned to use character tables to identify the representations to which the functions $x$, $y$, and $z$ belong. The first, and most precise, definition of the selection rule should therefore present no problems!

## 6.2 Raman spectroscopy

Raman spectroscopy involves the *emission* of light (*contrast* IR spectroscopy), in which the energies associated with vibrational transitions appear as *shifts in frequency* from a higher energy *exciting line*. A more complete description is again available elsewhere (see Appendix III), but the essential point to make in connection with symmetry is that a *symmetry selection rule* also exists for vibrational Raman spectroscopy.

Here, the rule may be stated as

*A vibration will be active in the Raman if it belongs to the same representation as one of the functions $x^2$, $y^2$, $z^2$, $xz$, $yz$ or $xy$.*

This definition also has the more qualitative analogue:

*A vibration will be active in the Raman if it involves a change in the polarisability of the molecule.*

In its precise form, this selection rule requires us to identify the representations of no less than six functions, and this information is also contained in the character table.

For *non-degenerate point groups*, these six functions are tabulated in the far right-hand column of the character table, and the corresponding representation is easily obtained. So, in the $C_{2v}$ point group, for example, $x^2$, $y^2$ and $z^2$ all transform as $A_1$ symmetry, and the functions $xz$, $yz$ and $xy$ have symmetries $B_1$, $B_2$ and $A_2$ respectively.

In point groups containing *degenerate* representations, these squared and cross-terms by themselves do not always act as bases for the irreducible representations. Rather, *linear combinations* of them must be used, and it is these linear combinations which are given in the character table. Raman activity is then associated with these linear combinations of the functions.

Thus, in the $C_{4v}$ point group, the functions $x^2$ and $y^2$ appear as the combinations $x^2 + y^2$ and $x^2 - y^2$. This latter combination indicates that $B_1$ modes will be Raman active. The functions $z^2$ and $x^2 + y^2$ separately impart Raman activity to $A_1$ modes. The $B_2$ and E modes will be Raman active by virtue of the cross-terms.

| $C_{2v}$ | E | $C_2$ | $\sigma_v(xz)$ | $\sigma_v'(yz)$ | $h = 4$ | | |
|---|---|---|---|---|---|---|---|
| $A_1$ | 1 | 1 | 1 | 1 | | $z$ | $x^2, y^2, z^2$ |
| $A_2$ | 1 | 1 | $-1$ | $-1$ | | $R_z$ | $xy$ |
| $B_1$ | 1 | $-1$ | 1 | $-1$ | | $x, R_y$ | $xz$ |
| $B_2$ | 1 | $-1$ | $-1$ | 1 | | $y, R_x$ | $yz$ |

| $C_{4v}$ | E | $2C_4$ | $C_2$ | $2\sigma_v$ | $2\sigma_d$ | $h = 8$ | | |
|---|---|---|---|---|---|---|---|---|
| $A_1$ | 1 | 1 | 1 | 1 | 1 | | $z$ | $x^2 + y^2, z^2$ |
| $A_2$ | 1 | 1 | 1 | $-1$ | $-1$ | | $R_z$ | |
| $B_1$ | 1 | $-1$ | 1 | 1 | $-1$ | | | $x^2 - y^2$ |
| $B_2$ | 1 | $-1$ | 1 | $-1$ | 1 | | | $xy$ |
| E | 2 | 0 | $-2$ | 0 | 0 | | $(R_x, R_y),$ $(x, y)$ | $(xz, yz)$ |

### 6.3    Mutual exclusion between IR and Raman spectral features

The different symmetry selection rules for IR and Raman spectroscopy can provide a useful method of identifying the symmetry of certain vibrations, and in *centrosymmetric molecules*, this distinction is particularly marked. For these molecules, it can be shown that *any vibration which is active in the IR will be inactive in the Raman, and that any vibration active in the Raman will be inactive in the IR.*

This mutual exclusion between IR and Raman spectral features arises here because of the fact that in centrosymmetric point groups, the simple functions $x$, $y$ and $z$ always belong to 'u'-type representations—i.e. those for which the symbol contains 'u' as a subscript, whilst the squared and cross-term functions which impart Raman activity all belong to 'g'-type representations.

It should be noted, however, that the observation of mutual exclusion does *not automatically imply* the existence of a centre of symmetry. There are a number of other point groups where mutual exclusion is *also to be found*, where the origin does not lie in the exclusivity of 'g' and 'u'. These include, for example, $D_{5h}$ (e.g. a pentagonal bipyramid or prism), $D_{4d}$ (a square antiprism) and $D_{6d}$.

In practice, however, the inference of a centre of symmetry from mutually exclusive IR and Raman data remains a very useful diagnostic indicator when distinguishing, for example, between *cis* and *trans* isomers, or between tetrahedral and square-planar configurations.

With these two selection rules, we are now in a position to comment on the IR and Raman activity of all the vibrations so far derived and, more importantly, to predict how these techniques might be used in the determination of molecular symmetry.

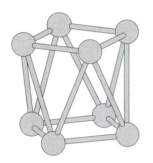

A square antiprism ($D_{4d}$)

### 6.4    IR and Raman active vibrations in $H_2O$ and $SO_2F_2$

Earlier in this primer, we derived the symmetries of the vibrations in $H_2O$ and $SO_2F_2$ to be

$$H_2O: \quad \Gamma_{vib} = 2A_1 + B_1, \qquad SO_2F_2: \quad \Gamma_{vib} = 4A_1 + A_2 + 2B_1 + 2B_2$$

Both molecules belong to the $C_{2v}$ point group, and in predicting which of these vibrations will be IR or Raman active, it is necessary to identify the representations to which the various functions ($x$, $x^2$, etc.) belong using the appropriate character table.

| $C_{2v}$ | E | $C_2$ | $\sigma_v(xz)$ | $\sigma_v'(yz)$ | $h = 4$ | |
|---|---|---|---|---|---|---|
| $A_1$ | 1 | 1 | 1 | 1 | $z$ | $x^2, y^2, z^2$ |
| $A_2$ | 1 | 1 | -1 | -1 | $R_z$ | $xy$ |
| $B_1$ | 1 | -1 | 1 | -1 | $x, R_y$ | $xz$ |
| $B_2$ | 1 | -1 | -1 | 1 | $y, R_x$ | $yz$ |

*IR activity*. Here we need to focus on $x$, $y$ and $z$, and from this, it may be deduced that only $A_1$, $B_1$ and $B_2$ vibrations will be IR active. For $H_2O$, all three vibrations should be present in the IR spectrum, but for $SO_2F_2$ only eight of the possible nine vibrations should be seen: the $A_2$ mode should be absent.

If only stretching modes are considered, we have

$$H_2O: \quad \Gamma_{stretch} = A_1 + B_1, \qquad SO_2F_2: \quad \Gamma_{stretch} = 2A_1 + B_1 + B_2$$

For both molecules, all stretching modes are therefore IR active.

*Raman activity*. This is associated with the functions $x^2$, $y^2$, $z^2$, $xz$, $yz$ and $xy$. The character table now shows that the first three functions all belong to the $A_1$ representation, and that the functions $xz$, $yz$ and $xy$ transform as $B_1$, $B_2$ and $A_2$ respectively. For both $H_2O$ and $SO_2F_2$, all vibrations are therefore Raman active. Indeed, for *any* $C_{2v}$ molecule, the Raman spectrum should always show *all* the vibrational fundamentals. For $SO_2F_2$, it should therefore be possible to identify the frequency of the single $A_2$ mode by comparing the IR and Raman vibrational data. In general, each mode is associated with a unique frequency, and a vibrational feature *observed in the Raman* but *absent in the IR* will correspond to the $A_2$ vibration.

## 6.5 Vibrations of molecules with higher symmetry— degenerate modes

We have already seen that the character tables of higher order point groups (e.g. $D_{4h}$, $T_d$) contain *degenerate representations*, i.e. representations for which the simplest matrices which obey the multiplication table are of order $2 \times 2$ or higher. When we come to study the vibrations of molecules in such point groups, we encounter *degenerate vibrations*—sometimes called *degenerate modes*. The $D_{4h}$ character table is reproduced below:

| $D_{4h}$ | $E$ | $2C_4$ | $C_2$ | $2C_2'$ | $2C_2''$ | $i$ | $2S_4$ | $\sigma_h$ | $2\sigma_v$ | $2\sigma_d$ | $h = 16$ | |
|---|---|---|---|---|---|---|---|---|---|---|---|---|
| $A_{1g}$ | 1 | 1 | 1 | 1 | 1 | 1 | 1 | 1 | 1 | 1 | | $x^2 + y^2, z^2$ |
| $A_{2g}$ | 1 | 1 | 1 | −1 | −1 | 1 | 1 | 1 | −1 | −1 | $R_z$ | |
| $B_{1g}$ | 1 | −1 | 1 | 1 | −1 | 1 | −1 | 1 | 1 | −1 | | $x^2 - y^2$ |
| $B_{2g}$ | 1 | −1 | 1 | −1 | 1 | 1 | −1 | 1 | −1 | 1 | | $xy$ |
| $E_g$ | 2 | 0 | −2 | 0 | 0 | 2 | 0 | −2 | 0 | 0 | $(R_x, R_y)$ | $(xz, yz)$ |
| $A_{1u}$ | 1 | 1 | 1 | 1 | 1 | −1 | −1 | −1 | −1 | −1 | | |
| $A_{2u}$ | 1 | 1 | 1 | −1 | −1 | −1 | −1 | −1 | 1 | 1 | $z$ | |
| $B_{1u}$ | 1 | −1 | 1 | 1 | −1 | −1 | 1 | −1 | −1 | 1 | | |
| $B_{2u}$ | 1 | −1 | 1 | −1 | 1 | −1 | 1 | −1 | 1 | −1 | | |
| $E_u$ | 2 | 0 | −2 | 0 | 0 | −2 | 0 | 2 | 0 | 0 | $(x, y)$ | |

### 6.6 Vibrations of XeF₄(D₄ₕ)

The symmetries of the vibrations in the square-planar molecule $XeF_4$ may be obtained in the usual way by subtracting $\Gamma_{trans}$ and $\Gamma_{rot}$ from $\Gamma_{mol}$. Using the unshifted atoms approach, the characters of $\Gamma_{mol}$ can be derived as:

|  | E | $2C_4$ | $C_2$ | $2C_2'$ | $2C_2''$ | i | $2S_4$ | $\sigma_h$ | $2\sigma_v$ | $2\sigma_d$ |
|---|---|---|---|---|---|---|---|---|---|---|
| No. of atoms unshifted | 5 | 1 | 1 | 3 | 1 | 1 | 1 | 5 | 3 | 1 |
| Contribution per atom (Table 5.1) | 3 | 1 | −1 | −1 | −1 | −3 | −1 | 1 | 1 | 1 |
| $\Gamma_{mol}$ | 15 | 1 | −1 | −3 | −1 | −3 | −1 | 5 | 3 | 1 |

From this the reduction formula gives

$$\Gamma_{mol} = A_{1g} + A_{2g} + 2A_{2u} + B_{1g} + B_{2g} + B_{2u} + E_g + 3E_u$$

Subtracting $\Gamma_{trans}$ and $\Gamma_{rot}$ then leads to

$$\Gamma_{vib} = A_{1g} + A_{2u} + B_{1g} + B_{2g} + B_{2u} + 2E_u$$

$$n = \frac{1}{h}\sum N_R \chi_R \chi_I$$

This list of vibrational symmetries gives little indication of what is happening in terms of atomic displacements, but it does tell us that we can expect the $A_{2u}$ and $E_u$ vibrations to be active in the IR (look for $x$, $y$, and $z$), and for the $A_{1g}$, $B_{1g}$, and $B_{2g}$ modes to be Raman active (look for squared terms and cross-terms). In addition, we can deduce that because each vibration is generally associated with a distinct frequency, there will be *mutual exclusion* between the IR and Raman spectra: i.e. a frequency observed in the IR will not be seen in the Raman, and vice versa. A better picture of these vibrations comes from considering the stretching and bending modes separately.

**Stretching modes in XeF₄**

The derivation of $\Gamma_{stretch}$ follows the general recipe described previously. After labelling the bonds, and noting the position of the various axes and planes in $XeF_4$, $\Gamma_{stretch}$ is derived by considering the number of unshifted bonds for each symmetry operation in $D_{4h}$. Figure 6.1 defines the axis system, and the characters may then be shown to be

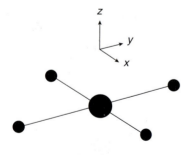

**Fig. 6.1**

|  | E | $2C_4$ | $C_2$ | $2C_2'$ | $2C_2''$ | i | $2S_4$ | $\sigma_h$ | $2\sigma_v$ | $2\sigma_d$ |
|---|---|---|---|---|---|---|---|---|---|---|
| $\Gamma_{stretch}$ | 4 | 0 | 0 | 2 | 0 | 0 | 0 | 4 | 2 | 0 |

which reduces to $\Gamma_{stretch} = A_{1g} + B_{1g} + E_u$. The representations $A_{1g}$ and $B_{1g}$ are composed of $1 \times 1$ matrices, and are termed *singly degenerate* or *non-degenerate* representations. As a consequence, the $A_{1g}$ and $B_{1g}$ stretching vibrations are described as being *singly degenerate* or *non-degenerate*.

The $A_{1g}$ *vibration* is the totally symmetric stretch, and corresponds to the simultaneous stretch or contraction of all four bonds, as shown in Fig. 6.2. This motion preserves all the symmetry elements in the molecule and, in particular, the *centre of symmetry*. This feature of the vibration is denoted by the subscript 'g' in the representation (an abbreviation for 'gerade').

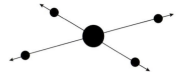

**Fig. 6.2** $A_{1g}$ stretch.

The $B_{1g}$ *vibration* is shown in Fig. 6.3. This vibration is also *centrosymmetric* (indicated by g), and although reflections in the planes $\sigma_v$ and in $\sigma_h$ are retained during the vibration, four-fold symmetry is destroyed.

In general, information concerning which symmetry elements are *retained* during a vibration is usefully summarised in the character of the corresponding representation. For singly degenerate vibrations such as those described above, if the vibration *retains* a particular symmetry element during its cycle, *the character is $+1$*, and if the symmetry element is *destroyed, the character is $-1$*.

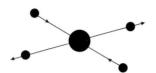

**Fig. 6.3** $B_{1g}$ stretch.

The $E_u$ *vibration* is a *doubly degenerate* stretch. In Chapter 4 we saw that the symbol E in a representation implies *two-fold degeneracy*, and the $E_u$ stretching vibration in $XeF_4$ is correspondingly termed a *doubly degenerate vibration*. The subscript 'u' in this symbol stands for 'ungerade', and refers to the fact that during this vibration, the centre of symmetry is lost.

The functions $x$ and $y$ taken together form a suitable basis for illustrating this representation (as may be seen from the character table), and the $E_u$ stretching mode can similarly be visualised as *two mutually perpendicular vibrations taking place along the x and y axes*. Figure 6.4 shows the changes in shape which correspond to the *two components* of this mode.

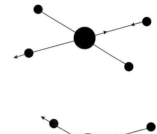

**Fig. 6.4** $E_u$ stretches.

The vibrational frequencies of these two components are *identical*, and only *one distinct* absorption would be observed in the IR. However, in terms of contributing to the total $(3n - 6)$ number of molecular vibrations, this $E_u$ mode contributes *two* vibrational degrees of freedom.

Finally, although Fig. 6.4 gives a satisfactory picture of the two components of this $E_u$ stretch, *it is not unique*, and other *equally valid* pictures of this mode can be derived. Unlike non-degenerate modes, where the atom displacements can in principle always be defined, for degenerate modes this is no longer true. The reason for this can be found in more advanced texts (Appendix III).

### Bending modes in $XeF_4$

The symmetries of the bending modes in $XeF_4$ may be obtained via $\Gamma_{bend} = \Gamma_{vib} - \Gamma_{stretch}$, from which

$$\Gamma_{bend} = A_{2u} + B_{2g} + B_{2u} + E_u$$

As was found above, the atom displacements for the *non-degenerate* modes can be uniquely visualised. Figure 6.5 shows these respective modes in pictorial form, and demonstrates the 'u'- and 'g'-type behaviour of these bending modes. Two of these modes can be seen to involve 'out-of-plane'

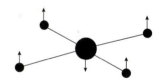

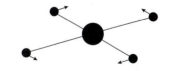

**Fig. 6.5** $A_{2u}, B_{2g}$ and $B_{2u}$ bends.

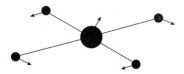

**Fig. 6.6** $E_u$ bend.

motion, and for the $A_{2u}$ mode in particular, this motion clearly belongs to the same representation as a displacement along the Cartesian '$z$' axis.

The $E_u$ doubly degenerate bend involves motion in the $xy$ plane, and Fig. 6.6 provides an illustration of one of the two components.

## 6.7 Vibrations in XY₄ (T_d) and XY₆ (O_h)

The point groups $T_d$ and $O_h$ are sometimes referred to as 'cubic' point groups, and the corresponding character tables contain both doubly degenerate and triply degenerate representations. As a result, it may be anticipated that the vibrations of these molecules involve both doubly and triply degenerate modes. Typical examples of molecules belonging to these point groups are $CH_4$ and $SF_6$.

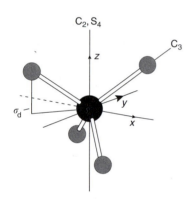

### $\Gamma_{vib}$, $\Gamma_{stretch}$ and $\Gamma_{bend}$ for CH₄

The positions of the symmetry elements in $CH_4$ are indicated in the figure alongside, and the first stage in deriving the various vibrational representations is to obtain the representation $\Gamma_{mol}$ using the unshifted atoms approach

|  | E | $8C_3$ | $3C_2$ | $6S_4$ | $6\sigma_d$ |
|---|---|---|---|---|---|
| No. of atoms unshifted | 5 | 2 | 1 | 1 | 3 |
| Contribution per atom (Table 5.1) | 3 | 0 | −1 | −1 | 1 |
| $\Gamma_{mol}$ | 15 | 0 | −1 | −1 | 3 |

This reduces to $\Gamma_{mol} = A_1 + E + T_1 + 3T_2$. Removing translations ($T_2$) and rotations ($T_1$) leaves $\Gamma_{vib} = A_1 + E + 2T_2$.

The effect of the symmetry operations on the four C–H bonds may be seen as

|  | E | $8C_3$ | $3C_2$ | $6S_4$ | $6\sigma_d$ |
|---|---|---|---|---|---|
| $\Gamma_{stretch}$ | 4 | 1 | 0 | 0 | 2 |

which reduces to $\Gamma_{stretch} = A_1 + T_2$, and the bending modes are therefore $\Gamma_{bend} = E + T_2$

As regards spectroscopic activity, only the $T_2$ modes are IR active, but all the modes are expected in the Raman. This molecule, although being highly symmetric, does not therefore show mutual exclusion, and the centre of symmetry is notably absent from the character table. The $A_1$ mode is totally symmetric, and the $T_2$ mode consists of three mutually orthogonal components, all of which have the same frequency. Figure 6.7 shows typical atom displacements for the component of the $T_2$ stretch along the $z$ axis.

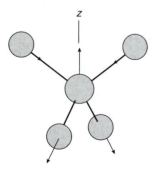

**Fig. 6.7**

## $\Gamma_{stretch}$ for SF$_6$

The molecule SF$_6$ belongs to the $O_h$ point group, and the symmetry operations in this structure have been identified (Chapter 1) as

$$E \quad 8C_3 \quad 6C_2' \quad 6C_4 \quad 3C_2(=C_4^2) \quad i \quad 6S_4 \quad 8S_6 \quad 3\sigma_h \quad 6\sigma_d$$

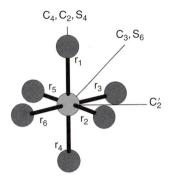

Fig. 6.8

The positions of the various axes are reproduced in Fig. 6.8. The three planes $\sigma_h$ are perpendicular to the three $C_4$ axes, whilst the planes $\sigma_d$ each contain a $C_4$ and $C_2'$ axis.

Following previous practice, the characters of the representation $\Gamma_{S-F}$ are obtained by noting the number of bonds unshifted during each of the symmetry operations. This leads to

| | E | 8C$_3$ | 6C$_2'$ | 6C$_4$ | 3C$_2$(= C$_4^2$) | i | 6S$_4$ | 8S$_6$ | 3σ$_h$ | 6σ$_d$ |
|---|---|---|---|---|---|---|---|---|---|---|
| $\Gamma_{S-F}$ | 6 | 0 | 0 | 2 | 2 | 0 | 0 | 0 | 4 | 2 |

The $O_h$ character table is reproduced in Appendix II and, using the reduction formula, the symmetries of the stretching modes may be derived as $\Gamma_{stretch} = A_{1g} + E_g + T_{1u}$.

$$n = \frac{1}{h}\sum N\chi_R\chi_I$$

As anticipated, there is one totally symmetric ($A_{1g}$) mode, and this corresponds to the in-phase stretching of all six bonds, as indicated in Fig. 6.9. The $E_g$ stretch consists of two components, one of which is shown in Fig. 6.10. In these modes, the centre of symmetry is preserved (subscript 'g'), and they are both Raman active by virtue of the combinations of square and cross-terms which appear in the final column of the character table. The $T_{1u}$ stretch comprises three components, one of which is shown in Fig. 6.11. This mode is IR active (look for $x$, $y$, or $z$), and is the only stretching mode in which the central atom moves.

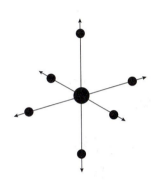

Fig. 6.9 $A_{1g}$ stretch.

## 6.8 Stretching modes in larger molecules: metal carbonyl species

As illustrated above, the use of internal coordinates as bases for the representations of stretching modes provides a means of isolating and extracting vibrational data for *specific sets of bonds*, and is particularly useful when those stretching modes are well removed in frequency from the other stretching or bending modes in the molecule.

This situation exists in many molecular species—for example, the C–H stretching modes in hydrocarbons—but it is perhaps in metal carbonyls and their derivatives where this simplification is most widespread and most useful. In these compounds, the C–O stretching vibrations ($\Gamma_{CO}$) are often well-separated from the other vibrations in the molecule, and also give intense absorptions in the IR spectrum. They are therefore easily identified.

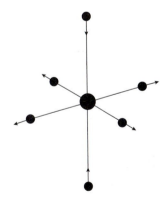

Fig. 6.10 $E_g$ stretch.

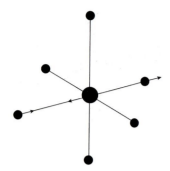

**Fig. 6.11** $T_{1u}$ stretch.

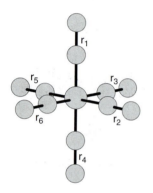

**Fig. 6.12**

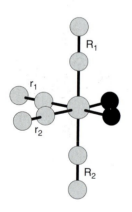

**Fig. 6.13**

In addition, the frequencies associated with particular vibrational symmetries *within* $\Gamma_{CO}$ are usually sufficiently separated such that they can be routinely distinguished from each other. Band-counting strategies based on predictions from symmetry therefore generally work satisfactorily and, as a result, the CO stretching region in carbonyl complexes is often used as a structural probe, offering the chance to establish the relative configuration of CO groups bound to a metal centre, and to distinguish between different isomers.

To illustrate this, we will focus on the symmetries of the C–O stretching modes in the species $Mo(CO)_6$ and in the typical derivatives *cis* and *trans* $Mo(CO)_4L_2$.

## 6.9 Carbonyl stretching modes in $Mo(CO)_6$

The symmetries of the C–O stretching modes in $Mo(CO)_6$ are obtained by considering the effect of the various symmetry operations on the six C–O bonds in the molecule (Fig. 6.12). However, comparison with the derivation of the stretching modes in $SF_6$ above shows that these two problems are *identical*: the C–O bonds in $Mo(CO)_6$ will behave in exactly the same way as the S–F bonds in $SF_6$ as regards the effect of the various symmetry operations.

The C–O stretching modes will therefore have symmetries

$$\Gamma_{C-O} = A_{1g} + E_g + T_{1u}$$

and, of these, $A_{1g}$ and $E_g$ will appear in the Raman spectrum and $T_{1u}$ in the IR. These modes are well-separated in frequency, lying at 2124, 2027, and 2004 cm$^{-1}$ respectively.

## 6.10 Carbonyl stretching modes in *cis* and *trans* $Mo(CO)_4L_2$

The *cis* complex $Mo(CO)_4L_2$ has $C_{2v}$ symmetry if the two ligands L can be regarded as single atom donors, and this configuration therefore contains *two sets of C–O bonds*. Two of the carbonyl groups are *trans* to ligands L, and are labelled as $r_1$ and $r_2$, whilst the other pair ($R_1$ and $R_2$) are *trans* to each other (Fig. 6.13).

When considering the stretching vibrations of the CO groups, the representation $\Gamma_{C-O}$ may therefore be treated in two separate parts: $\Gamma_{C-O} = \Gamma_R + \Gamma_r$. Using this simplification, it is relatively straightforward to show that $\Gamma_{C-O} = 2A_1 + B_1 + B_2$. However, it is important to recognise that this exercise is *identical* to that previously carried out for $SO_2F_2$, which similarly has two sets of bonds in orthogonal planes and also belongs to the $C_{2v}$ point group. For this complex, all four CO stretches are active in both the IR and Raman spectra.

The *trans* isomer of this complex is shown in Fig. 6.14. It has symmetry $D_{4h}$, and the derivation of $\Gamma_r$ would follow the normal procedure. However, this problem is again identical to one encountered previously—in this case, to the derivation of $\Gamma_{Xe-F}$ in $XeF_4$. Thus $\Gamma_{C-O} = A_{1g} + B_{1g} + E_u$, and there are two Raman active modes ($A_{1g} + B_{1g}$) and one IR active mode ($E_u$).

Here, the presence of the centre of symmetry results in mutual exclusion between IR and Raman, and it is clear that selection rules predict that this isomer should be distinguishable from the *cis* variety as a result of the different numbers of CO stretching bands observed in either the IR or the Raman spectrum.

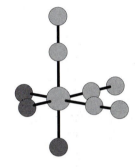

**Fig. 6.14**

## 6.11 Summary

This chapter has attempted to provided a rationale for the earlier discussions on point groups, symmetry operations, matrices and representations. In general, most spectroscopic properties of molecules can only be interpreted in relation to their symmetry, and, conversely, spectroscopic techniques and their selection rules are central in the determination of molecular shape.

It should be noted, however, that symmetry selection rules in general make no predictions about band intensities: only that a band is expected to be present or absent. They also make no allowance for accidental band overlap between modes of different symmetry. Both these factors should be borne in mind when dealing with real systems.

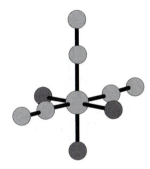

**Fig. 6.15** *fac* $ML_3(CO)_3$.

## 6.12 Exercises

1. Derive the symmetries of *all* the vibrations in the $NH_3$ molecule (i.e. $\Gamma_{vib}$). Identify which of these are stretching modes and which are bending modes, and state how many bands you would expect to observe in the IR spectrum of this molecule.
2. Derive the symmetries of the stretching modes in $PCl_3F_2$ (e.g. Fig. 1.25). Which of these modes will be observed in *both* the IR and the Raman spectrum?
3. How many stretching modes should be observed in the Raman spectrum of $C_2H_4$? (see Fig. 3.5).
4. The IR spectrum of molecular $XeO_3F_2$ shows one Xe–O stretch and one Xe–F stretch. Propose a shape for this molecule consistent with these observations.
5. One of the isomers of the ion $[PtCl_2Br_2]^{2-}$ has only two IR active stretching modes. Identify its point group.
6. Show how you might use IR spectroscopy in the CO stretching region to distinguish between (a) *cis* and *trans* $ML_4(CO)_2$, and (b) *fac* and *mer* $ML_3(CO)_3$ (see Figs 6.15 and 6.16).

**Fig. 6.16** *mer* $ML_3(CO)_3$.

# 7 Symmetry aspects of chemical bonding

The final chapter in this Primer is concerned with the use of symmetry in describing and depicting the chemical bonding and electronic energy levels within molecules. In the molecular orbital description of chemical bonding, the molecular wave functions which ultimately establish these energy levels have symmetry properties which relate to the point group of the molecule, such that, in general terms, each wave function must have a symmetry corresponding to one of the irreducible representations.

These wave functions are 'constructed' from linear combinations of the constituent atomic orbitals (the LCAO model) and in carrying out this step, it is helpful to introduce symmetry arguments at an early stage. At the same time, it is useful to retain the ideas of localised bonding which are inherent in the terms '$\sigma$-bond' and '$\pi$-bond'.

A comprehensive description of chemical bonding based on the LCAO model can be found in some of the texts listed in Appendix III. In this chapter, we shall focus primarily on deriving the symmetries of the orbital combinations which form the building blocks for this model.

## 7.1 Assemblies of atoms—symmetry description of orbital combinations

In Chapter 5, we saw how molecular vibrations arose as a consequence of the correlated motion of a group of atoms. Initially, with the atoms well-separated and non-interacting, as in an ideal monatomic gas, the $3n$ degrees of freedom of an n-atom system all correspond to translations, and the motion of the atoms is uncorrelated. As the atoms are brought together to form a well-defined grouping, only three translational degrees of freedom remain—those of the resulting molecule—and the remaining degrees of freedom appear as molecular rotations and vibrations. Vibrational symmetry then arises as a result of the wave functions needed to describe the vibrational energy levels in the molecule.

An analogous situation arises concerning the energies of assemblies of atoms. Figure 7.1 shows a random assembly of atoms, and for simplicity, these can be taken to be hydrogen atoms, each with a $1s$ orbital. The system as a whole is dynamic, has no defined equilibrium structure, and its energy is the sum of the energies of the individual non-interacting atoms.

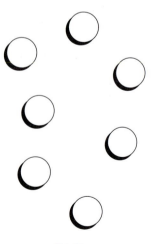

Fig. 7.1

## 7.2   Atomic orbitals in a tetrahedral ($T_d$) array

Imagine now a situation in which four such atoms interact such that their *relative equilibrium positions* remain fixed—for example, in a tetrahedral array—to produce the 'molecule' $H_4$. The energy levels in this structure now depend in part on the *interaction* between the individual atomic orbitals, and the wave functions which describe this are based on *combinations* of constituent orbitals. The initial aim in this chapter is to establish a symmetry description of such combinations.

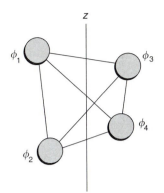

**Fig. 7.2**

Figure 7.2 shows four hydrogen atoms in a tetrahedral array, with individual atomic orbitals $\phi_1$, $\phi_2$, $\phi_3$ and $\phi_4$, and these must be combined in such a way as to correspond to irreducible representations in $T_d$—i.e. we require $\Gamma_\phi$.

This representation is obtained by considering the effect of the operations in the $T_d$ point group on the four orbitals in question. This problem is *identical* to an example previously discussed in Chapter 4, where we derived the representation for the $1s$ orbitals on the H atoms in $CH_4$ as

$$\Gamma_\phi = A_1 + T_2$$

When discussing the symmetries of molecular vibrations, we saw that each vibrational mode (e.g. $A_1$, $B_1$ etc.) had an identifiable frequency, and an analogous situation arises here.

The symmetry combinations $A_1 + T_2$ which arise here from the four constituent atomic orbitals in the assembly '$H_4$' similarly give rise to two distinct energy levels.

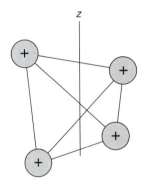

**Fig. 7.3**

Finally, just as it is possible to visualise the relative atomic displacements involved in a vibration of a particular symmetry, so it is possible to indicate the *relative signs* on the constituent atomic orbitals which combine to give $\Gamma_\phi$.

Figure 7.3 shows the $A_1$ combination of $1s$ orbitals in '$H_4$'. It is clearly a 'totally symmetric' array of positive signs. From what we have discussed earlier, the $T_2$ combination will have three components consistent with its designation as 'triply degenerate', and some similarity with the symmetries of the displacement coordinates $x$, $y$ and $z$ is to be anticipated, since these functions also transform as $T_2$. One component of the $T_2$ combination is shown in Fig. 7.4 and, as expected, the array of '+' and '−' signs on the orbitals mirrors the '+' and '−' directions on the related Cartesian axis—in this case, $z$.

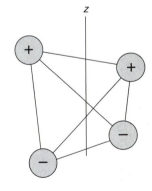

**Fig. 7.4**

## 7.3   Other arrays of s-orbitals

The representations $A_1 + T_2$ may also be recognised as the symmetries of the stretching modes in a $T_d$ molecule such as $CH_4$ (Chapter 6), and this correspondence between $\Gamma_\phi$ for an assembly of '$n$' $1s$ orbitals, and the stretching modes $\Gamma_{M-L}$ in a molecule $ML_n$ can usefully be extended to many other systems.

Thus an octahedral assembly of six s-orbitals will transform in the same way as $\Gamma_{S-F}$ in $SF_6$ ($A_{1g} + E_g + T_{1u}$), and a square-planar array of s-orbitals ($D_{4h}$) will have the representation $A_{1g} + B_{1g} + E_u$ by analogy with $\Gamma_{Xe-F}$ in $XeF_4$—both results having also been previously derived in Chapter 6.

## 7.4  Central Atom Orbitals

Although several molecules are known which comprise a simple cluster of identical atoms (e.g. $P_4$), many more molecules have a structure in which there is a 'central' atom coordinated by two or more ligands. This description clearly applies to $CH_4$, $SF_6$ and $XeF_4$, but would also extend to $NH_3$ and $H_2O$.

In these species, the molecular orbital bonding model involves an interaction between the ligand orbitals and the central atom orbitals which is dictated by symmetry.

In general, a central atom may possess orbitals of types s, p, d, etc., and for an isolated atom in free space these orbitals retain their respective one-fold, three-fold, five-fold, etc. degeneracies. This means that the set of p-orbitals ($p_x$, $p_y$ and $p_z$), for example, remain energetically equivalent, and the same would be true for the five d-orbitals. The environment of such an atom could be regarded as 'spherical', in that free space has no preferred direction.

However, if a central atom finds itself in a lower symmetry environment, produced for example by a set of ligands, there are two general consequences. Firstly, it becomes important to be aware of the symmetry representation(s) to which the orbitals now belong in their new environment, and secondly to recognise that the degeneracies of one or more orbital types may be lost.

### Central atom p-orbitals

In an isolated xenon atom, the valence orbitals $5p_x$, $5p_y$ and $5p_z$ are all energetically equivalent, and so triply degenerate. However, if this atom is placed in the centre of a square-planar array of fluorine atoms, as in $XeF_4$, the degeneracy is partially removed. The orbitals $5p_x$ and $5p_y$ remain equivalent, but different from the $5p_z$. This is clearly evident from Fig. 7.5, where there are fluorine atoms lying on the x and y axes, but none lying on z.

We can now *classify* the p-orbitals according to the representations to which they belong in $D_{4h}$ symmetry, and, remembering that the orbitals $p_x$, $p_y$ and $p_z$ have the same symmetry as the functions x, y and z, the character table (below) shows that the three 5p orbitals transform as $E_u + A_{2u}$ in $D_{4h}$.

This character table may also be used to illustrate how the 5s and 5d orbitals on the xenon atom transform in $D_{4h}$ symmetry, but to obtain these representations, we need to know the mathematical functions which correspond to the s- and d- orbitals.

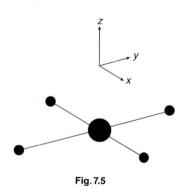

Fig. 7.5

| $D_{4h}$ | E | $2C_4$ | $C_2$ | $2C_2'$ | $2C_2''$ | i | $2S_4$ | $\sigma_h$ | $2\sigma_v$ | $2\sigma_d$ | $h = 16$ | |
|---|---|---|---|---|---|---|---|---|---|---|---|---|
| $A_{1g}$ | 1 | 1 | 1 | 1 | 1 | 1 | 1 | 1 | 1 | 1 | | $x^2 + y^2, z^2$ |
| $A_{2g}$ | 1 | 1 | 1 | $-1$ | $-1$ | 1 | 1 | 1 | $-1$ | $-1$ | $R_z$ | |
| $B_{1g}$ | 1 | $-1$ | 1 | 1 | $-1$ | 1 | $-1$ | 1 | 1 | $-1$ | | $x^2 - y^2$ |
| $B_{2g}$ | 1 | $-1$ | 1 | $-1$ | 1 | 1 | $-1$ | 1 | $-1$ | 1 | | $xy$ |
| $E_g$ | 2 | 0 | $-2$ | 0 | 0 | 2 | 0 | $-2$ | 0 | 0 | $(R_x, R_y)$ | $(xz, yz)$ |
| $A_{1u}$ | 1 | 1 | 1 | 1 | 1 | $-1$ | $-1$ | $-1$ | $-1$ | $-1$ | | |
| $A_{2u}$ | 1 | 1 | 1 | $-1$ | $-1$ | $-1$ | $-1$ | $-1$ | 1 | 1 | $z$ | |
| $B_{1u}$ | 1 | $-1$ | 1 | 1 | $-1$ | $-1$ | 1 | $-1$ | $-1$ | 1 | | |
| $B_{2u}$ | 1 | $-1$ | 1 | $-1$ | 1 | $-1$ | 1 | $-1$ | 1 | $-1$ | | |
| $E_u$ | 2 | 0 | $-2$ | 0 | 0 | $-2$ | 0 | 2 | 0 | 0 | $(x, y)$ | |

## Central atom s-orbitals

All *s*-orbitals are spherically symmetric, differing only in the numbers of radial nodes. They are non-degenerate, and as the equation which generates a sphere (in Cartesian coordinates) is $x^2 + y^2 + z^2 = r^2$ (a constant), we should look for the above function in the $D_{4h}$ character table. This particular character table does not give this function explicitly, but rather lists $(x^2 + y^2)$ and $z^2$ separately as each having $A_{1g}$ symmetry. This is sufficient to ensure that the *sum* of these two functions $[(x^2 + y^2) + z^2]$ also has $A_{1g}$ symmetry, and *s*-orbitals in $D_{4h}$ therefore transform as $A_{1g}$.

## Central atom d-orbitals

Figure 7.6 shows the familiar shapes associated with the five *d*-orbitals. The mathematical functions which describe the shapes of the five *d*-orbitals are $(2z^2 - x^2 - y^2)$, $(x^2 - y^2)$, $xy$, $xz$ and $yz$, and it is the signs of these functions within the coordinate system that are responsible for the patterns of '+' and '−' signs on the various lobes in these orbitals.

The last four of these functions form the familiar subscripts which identify orbitals such as $d_{xy}$ or $d_{yz}$, but the function $(2z^2 - x^2 - y^2)$ may be unfamiliar. It is in fact the correct subscript for the *d*-orbital more generally known as $d_{z^2}$. In many point groups, the function $z^2$ has the *same symmetry*—i.e. belongs to the same representation—as the function $(2z^2 - x^2 - y^2)$, and the orbital description may then be shortened to '$z^2$' for convenience.

In some higher symmetry point groups (notably $O_h$ and $T_d$), the full subscript is necessary, and is listed as such in the character table, but the majority of character tables just list $z^2$—as in $D_{4h}$.

However, the main conclusion which can be drawn here is that for a central atom in a $D_{4h}$ environment, much of the five-fold degeneracy is lost, with the above *d*-orbital functions now transforming according to the $A_{1g}$, $B_{1g}$, $B_{2g}$ and $E_g$ representations.

## Note on orbital labelling

The representations which appear in the various character tables are identified by a labelling system which incorporates upper case letters (or

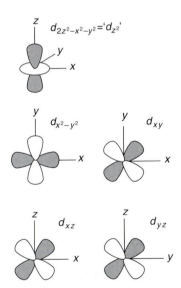

**Fig. 7.6**

Greek symbols, in the case of linear molecules). However, when labelling *atomic orbitals*—on diagrams, for example—it is conventional to use *lower case lettering*, rather than upper case.

Thus the individual *d*-orbitals in a $D_{4h}$ environment would be *labelled* as $a_{1g} + b_{1g} + b_{2g} + e_g$, and an *s*-orbital as $a_{1g}$.

Two quite general conclusions may be drawn from this:

1.  Central atom *s*-orbitals always transform according to the totally symmetric representation.
2.  Other central atom orbitals (e.g. *p*- or *d*-) transform in the same way as the mathematical functions which appear as their subscripts—whilst noting the simplification made in the case of $2z^2 - x^2 - y^2$.

## 7.5  The σ-framework in CH$_4$

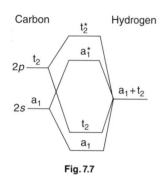

**Fig. 7.7**

We are now in a position to construct a molecular orbital diagram for the bonding in $CH_4$. This is done by combining orbitals on the central carbon atom with ligand orbital combinations of the same symmetry, to produce bonding and antibonding combinations. The orbitals on carbon are the valence 2*s* and 2*p* set and, in a tetrahedral environment, these transform as $a_1$ and $t_2$ respectively. The set of four hydrogen orbitals also transforms as $a_1 + t_2$, and the molecular orbital diagram which describes the effect of combining these sets of atomic orbitals is shown in Fig. 7.7. In methane there are a total of eight valence electrons, four from the carbon atom, and one each from the hydrogen atoms, and this is just the right number to fill the $a_1 + t_2$ bonding levels.

## 7.6  Combinations of *p*$_σ$ orbitals: σ-bonds in XeF$_4$

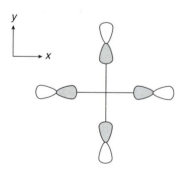

**Fig. 7.8**

The above example shows how a set of *s*-orbitals can interact with a central atom to form a σ-framework, but *p*-orbitals *with their lobes lying along the internuclear axis* can also participate in σ-bonding, as in this orientation, they are *axially symmetric*.

Figure 7.8 shows the four *p*-orbitals on the fluorine atoms in $XeF_4$ ($D_{4h}$) which lie along the Xe–F axes. In this orientation, they are behaving just like *s*-orbitals, and their representation $\Gamma_p$ in $D_{4h}$ is therefore *identical to that for the stretching modes* in $XeF_4$, derived previously: $\Gamma_p = a_{1g} + b_{1g} + e_u$.

The central xenon atom has orbitals of matching symmetry in the 5*s* ($a_{1g}$), $5d_{x^2-y^2}$ ($b_{1g}$) and the pair $5p_x$, $5p_y$ ($e_u$) and, just as in the case of $CH_4$ above, a molecular orbital diagram can be constructed in which xenon and fluorine orbitals of the same symmetry interact to give bonding and antibonding combinations. Figure 7.9 shows the bonding combinations of xenon orbitals and fluorine *p*-orbitals in the σ-system for $XeF_4$.

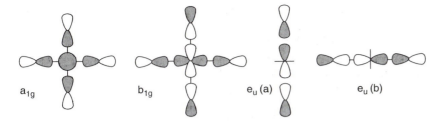

**Fig. 7.9** $\sigma$-bonding in $XeF_4$.

## 7.7 Out-of-plane $\pi$-bonding in $XeF_4$

Figure 7.10 shows the orientation of $p_z$ orbitals on the four fluorine atoms in $XeF_4$, and they clearly offer the possibility of out-of-plane $\pi$-bonding with respect to a single Xe–F bond. However, in order to establish the symmetries of any possible *molecular* $\pi$-bonds, we must obtain the irreducible representations for $\Gamma_\pi$ in the $D_{4h}$ point group.

The characters of the matrices which make up $\Gamma_\pi$ are obtained in the usual way by considering the effect of the various symmetry operations on the four $p_z$ orbitals, whilst noting that these out-of-plane fluorine orbitals can now be *reversed* by a symmetry operation such as $\sigma_h$. Using the molecular orientation adopted previously (Fig. 1.12), we have

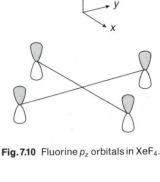

**Fig. 7.10** Fluorine $p_z$ orbitals in $XeF_4$.

| | E | $2C_4$ | $C_2$ | $2C_2'$ | $2C_2''$ | i | $2S_4$ | $\sigma_h$ | $2\sigma_v$ | $2\sigma_d$ |
|---|---|---|---|---|---|---|---|---|---|---|
| $\Gamma_\pi$ | 4 | 0 | 0 | $-2$ | 0 | 0 | 0 | $-4$ | 2 | 0 |

This reduces to $\Gamma_\pi = a_{2u} + b_{2u} + e_g$.

The $D_{4h}$ character table shows that the orbitals on the central xenon atom which have the correct symmetries to interact with these fluorine $p_z$ combinations are the $5p_z$ ($a_{2u}$), and the pair of orbitals $5d_{xz}$ and $5d_{yz}$, which together transform as $e_g$. There is no orbital on the xenon atom with $b_{2u}$ symmetry. Figure 7.11 shows the $a_{2u}$ and $e_g$ combinations of atomic

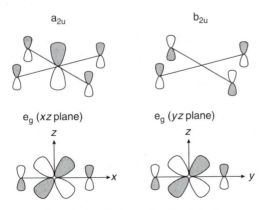

**Fig. 7.11** Out-of-plane $\pi$-bonding in $XeF_4$.

orbitals which result in net bonding, and also the non-bonding $b_{2u}$ combination of F orbitals.

### 7.8 Ligand $\sigma$- and $\pi$-orbitals in octahedral $MX_6$ complexes

Finally in this chapter, we shall look at the possibilities for $\sigma$- and $\pi$-bonding in transition metal octahedral complexes from the viewpoint of symmetry. As in the example above, the ligand $\pi$-orbitals will again generally be $p$-orbitals perpendicular to the $\sigma$-framework, but as the central atom $s$- and $p$-orbitals are already totally committed in the $\sigma$-system, we can anticipate that any $\pi$-bonding must involve the metal $d$-orbitals.

The starting point here is to recognise that each ligand X generally has available one $p$-orbital of $\sigma$-symmetry, which is directed towards the central atom M, and which will contribute to the $\sigma$-framework, and two further $p$-orbitals perpendicular to the M–X bond, which are available for $\pi$-bonding. The orientations of these $p$-orbitals with respect to the M–X bond are shown in Fig. 7.12. In total, there are six ligands X coordinated to M, and hence there are six ligand $p$-orbitals which can participate in $\sigma$-bonding, and 12 ligand $p$-orbitals in a possible $\pi$-system.

The representation for ligand $\sigma$-bonding, $\Gamma_\sigma$, can be recognised as being identical to $\Gamma_{stretch}$ for $MX_6$. This has previously been derived as $\Gamma_{stretch} = A_{1g} + E_g + T_{1u}$ (Chapter 6), and when applying this result to the $\sigma$-orbitals, we need only to remember to use lower case labels. Hence $\Gamma_\sigma = a_{1g} + e_g + t_{1u}$. The central atom can provide orbitals with these symmetries, and Fig. 7.13 depicts the $\sigma$-bonding combinations of ligand and metal orbitals.

The representation $\Gamma_\pi$ may be obtained by considering the effect of the various symmetry operations in the $O_h$ point group on the 12 ligand $p$–$\pi$ orbitals, and then reducing the resulting representation using the reduction formula. This procedure works satisfactorily, but there is an

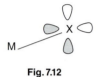

**Fig. 7.12**

$$n = \frac{1}{h} \sum N\chi_R\chi_I$$

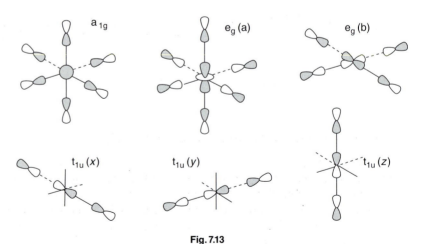

**Fig. 7.13**

alternative and quicker method which uses the fact that atomic $p$-orbitals transform in the same way as the atomic displacements $x$, $y$ and $z$. Hence, for each ligand X,

$$\Gamma_p = \Gamma_{mol}$$

If we define $\Gamma_X$ as the representation for *all 18* X-ligand $p$-orbitals, then

$$\Gamma_X = \Gamma_\sigma + \Gamma_\pi$$

However, because of the equivalence between $p$-orbitals and $x$, $y$ and $z$, $\Gamma_X$ *is equivalent to working out* $\Gamma_{mol}$ *for the set of six ligand atoms* X.

We therefore need only to calculate $\Gamma_{mol}$ for the six ligand atoms, and then obtain $\Gamma_\pi$ from $\Gamma_\pi = \Gamma_{mol} - \Gamma_\sigma$.

Using the procedure described in Chapter 5, $\Gamma_{mol}$ is obtained from a knowledge of the *number of unshifted atoms*, and the *contributions per unshifted* atom (Table 5.1) as follows:

| | E | $8C_3$ | $6C_2$ | $6C_4$ | $3C_2(= C_4^2)$ | i | $6S_4$ | $8S_6$ | $3\sigma_h$ | $6\sigma_d$ |
|---|---|---|---|---|---|---|---|---|---|---|
| No. of unshifted atoms | 6 | 0 | 0 | 2 | 2 | 0 | 0 | 0 | 4 | 2 |
| Contribution per atom | 3 | 0 | −1 | 1 | −1 | −3 | −1 | 0 | 1 | 1 |
| $\Gamma_{mol}$ | 18 | 0 | 0 | 2 | −2 | 0 | 0 | 0 | 4 | 2 |

This may be reduced using the reduction formula to give

$$\Gamma_{mol} = a_{1g} + e_g + t_{1g} + t_{2g} + 2t_{1u} + t_{2u}$$

The representation $\Gamma_\sigma$ has been derived as $a_{1g} + e_g + t_{1u}$, and the symmetry combination for the 12 ligand orbitals which can participate in $\pi$-bonding is therefore

$$\Gamma_\pi = t_{1g} + t_{2g} + t_{1u} + t_{2u}$$

## 7.9 Molecular orbital bonding scheme for MX$_6$

The symmetry properties of the central metal $s$-, $p$- and $d$-orbitals are obtained from the $O_h$ character table as described previously. From this table, it may be deduced that they transform as $a_{1g}$, $t_{1u}$ and $(e_g + t_{2g})$ respectively, and with this information it is now possible to construct the molecular orbital diagram for MX$_6$. Figure 7.14 shows how this can ultimately be accomplished.

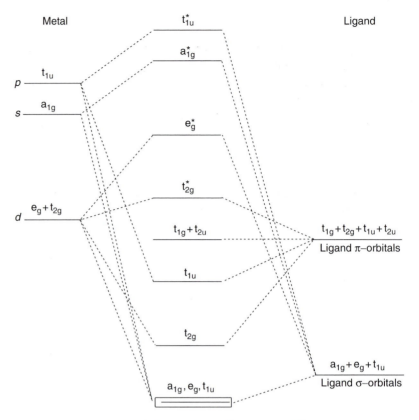

**Fig. 7.14** Schematic diagram for bonding in $MX_6$.

## 7.10   Summary

The principal aim of this chapter has been to demonstrate how to set up molecular orbital bonding schemes by considering the symmetries of participating sets of atomic orbitals. However, although it is a general result that each identifiable molecular orbital has a distinct energy, symmetry considerations do not provide any firm quantitative information regarding the energies of these orbitals. In this respect, the situation is very similar to vibrational spectroscopy.

What has been evident, however, is that the same representations have been turning up in this bonding chapter as in the previous chapters describing molecular vibrations. This common thread is one of the more interesting and useful aspects of symmetry, and one of the main goals of this Primer has been to demonstrate how widespread an approach based on symmetry can be.

The final exercises below are designed to consolidate the material in this chapter.

### 7.11   Exercises

1.   The molecule $BF_3$ has $D_{3h}$ symmetry. Derive the symmetries of the F atom $p$-orbitals involved in $\sigma$-bonding in this molecule, and state which atomic orbital(s) on the central boron atom could be used in $\sigma$-bonding.

2.   Derive the symmetries of the F atom $p$-orbitals suitable for out-of-plane $\pi$-bonding in $BF_3$. Which boron orbital could be involved in such $\pi$-bonding?

3.   Show that the symmetries of the *ligand* $\pi$-orbitals in a tetrahedral metal complex $ML_4$ transform as $\Gamma_L = e + t_1 + t_2$. Which $d$-orbitals on the central metal atom could participate in $\pi$-bonding in such a complex?

4.   Alkali metal atom clusters of stoichiometry $M_8$ have recently been identified, and one possible shape for these is a square antiprism (see the figure alongside). Assuming that each alkali metal provides one $s$-orbital ($\phi$) for cluster formation, derive the representation $\Gamma_\phi$ for a cluster of this shape. (The point group of this cluster appears in Chapter 6.)

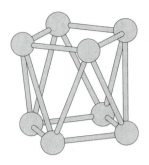

Possible shape of $M_8$ cluster

# Appendix I

## Answers to exercises

### Chapter 1

1. (a) $D_{3h}$, (b) $C_{4v}$, (c) $C_{3v}$, (d) $C_{2h}$, (e) $D_{6h}$, (f) $C_{2v}$, (g) $C_{2v}$, (h) $D_{3h}$, (i) $C_{3v}$, (j) $D_{5h}$, (k) $D_{5d}$.
2. Structures (d), (e) and (k) have a centre of inversion. Rotation–reflection axes occur in (a) $S_3$, (e) $S_3$ *and* $S_6$, (h) $S_3$, (j) $S_5$, (k) $S_{10}$.
3. A propellor with four blades has this symmetry (cf. Fig. 1.29).
4. The most straightforward shapes would be those based on planar objects with a single nine-fold axis—but care must be taken to ensure that no perpendicular $C_2$ axes are present, which would result in $D_{9h}$. Lowering the symmetry to $C_9$ may then be achieved by placing an extra object on the $C_9$ axis above the original plane (cf. Fig. 1.14).

### Chapter 2

1. (i) 14, (ii) 2, (iii) 2, (iii) 19.
2. (i) $\sigma(xz)$, (ii) $C_2(y)$, (iii) $\sigma(xy)$.
3. The functions $x$, $y^2$ and $xz$ belong to the representations $B_{3u}$, $A_g$ and $B_{2g}$.
4. (i) $x^2$, $y^2$ or $z^2$ would all be possible functions. (ii) $z$. (iii) $xy$.

### Chapter 3

1. $\Gamma_1 = A_1 + 2A_2 + B_1$; $\Gamma_2 = A_1 + A_2 + B_1 + B_2$.
2. (i) $A_g + B_u$; (ii) $3A_1 + 2B_1$; (iii) $A_g + B_{3g} + B_{1u} + B_{2u}$.

### Chapter 4

1. $\Gamma_{N-H} = A_1 + E$; $\Gamma_{B-F} = A_1' + E'$; $\Gamma_{Cl-F} = 2A_1 + B_1$.
2. $\Gamma_{Xe-O} = A_1$; $\Gamma_{Xe-F} = A_1 + B_1 + E$.
3. $\Gamma_{B-Cl} = A_1 + B_1 + E$; $\Gamma_{B-B} = A_1$.

### Chapter 6

1. $\Gamma_{vib} = 2A_1 + 2E$, $\Gamma_{stretch} = A_1 + E$ and $\Gamma_{bend} = A_1 + E$. Four bands would be expected in the IR spectrum.
2. $\Gamma_{P-Cl} = A_1' + E'$, $\Gamma_{P-F} = A_1' + A_2''$. The $E'$ mode will appear in both.
3. Three stretching modes should be observed: $2A_g + B_{3g}$.
4. Trigonal bipyramid ($D_{3h}$).
5. $D_{2h}$ (*trans* planar).
6. (a) The *cis* isomer ($C_{2v}$) should show two CO stretches in the IR; the *trans* ($D_{4h}$) should only show one. (b) The *fac* isomer ($C_{3v}$) should show two CO stretches in the IR; the *mer* should show three.

**Chapter 7**

1. $\Gamma_\sigma = a_1' + e'$. The boron orbitals would be the $2s$ and the pair $2p_x$, $2p_y$.
2. $\Gamma_\pi = a_2'' + e''$. The boron $2p_z$ orbital (symmetry $a_2''$) could be involved.
3. The $d$-orbitals transform as $e + t_2$ in $T_d$. All five could participate.
4. The square antiprism belongs to the point group $D_{4d}$. By considering the number of unshifted orbitals, the characters of $\Gamma_\phi$ turn out to be

| | E | $2S_8$ | $2C_4$ | $2S_8^3$ | $C_2$ | $4C_2'$ | $4\sigma_d$ |
|---|---|---|---|---|---|---|---|
| $\Gamma_\phi$ | 8 | 0 | 0 | 0 | 0 | 0 | 2 |

from which $\Gamma_\phi = A_1 + B_2 + E_1 + E_2 + E_3$. The molecular orbitals for the $M_8$ cluster would therefore have these symmetries.

# Appendix II

## Character tables for selected point groups

| $C_2$ | E | $C_2$ | $h = 2$ | |
|---|---|---|---|---|
| A | 1 | 1 | $z, R_z$ | $x^2, y^2, z^2, xy$ |
| B | 1 | $-1$ | $x, y, R_x, R_y$ | $yz, xz$ |

| $D_2$ | E | $C_2(z)$ | $C_2(y)$ | $C_2(x)$ | $h = 4$ | |
|---|---|---|---|---|---|---|
| A | 1 | 1 | 1 | 1 | | $x^2, y^2, z^2$ |
| $B_1$ | 1 | 1 | $-1$ | $-1$ | $z, R_z$ | $xy$ |
| $B_2$ | 1 | $-1$ | 1 | $-1$ | $y, R_y$ | $xz$ |
| $B_3$ | 1 | $-1$ | $-1$ | 1 | $x, R_x$ | $yz$ |

| $D_3$ | E | $2C_3$ | $3C_2$ | $h = 6$ | |
|---|---|---|---|---|---|
| $A_1$ | 1 | 1 | 1 | | $x^2 + y^2, z^2$ |
| $A_2$ | 1 | 1 | $-1$ | $z, R_z$ | |
| E | 2 | $-1$ | 0 | $(x, y), (R_x, R_y)$ | $(x^2 - y^2, xy), (xz, yz)$ |

| $D_4$ | E | $2C_4$ | $C_2(= C_4^2)$ | $2C_2'$ | $2C_2''$ | $h = 8$ | |
|---|---|---|---|---|---|---|---|
| $A_1$ | 1 | 1 | 1 | 1 | 1 | | $x^2 + y^2, z^2$ |
| $A_2$ | 1 | 1 | 1 | $-1$ | $-1$ | $z, R_z$ | |
| $B_1$ | 1 | $-1$ | 1 | 1 | $-1$ | | $x^2 - y^2$ |
| $B_2$ | 1 | $-1$ | 1 | $-1$ | 1 | | $xy$ |
| E | 2 | 0 | $-2$ | 0 | 0 | $(x, y), (R_x, R_y)$ | $(xz, yz)$ |

| $C_{2v}$ | E | $C_2$ | $\sigma_v(xz)$ | $\sigma_v'(yz)$ | $h = 4$ | |
|---|---|---|---|---|---|---|
| $A_1$ | 1 | 1 | 1 | 1 | $z$ | $x^2, y^2, z^2$ |
| $A_2$ | 1 | 1 | $-1$ | $-1$ | $R_z$ | $xy$ |
| $B_1$ | 1 | $-1$ | 1 | $-1$ | $x, R_y$ | $xz$ |
| $B_2$ | 1 | $-1$ | $-1$ | 1 | $y, R_x$ | $yz$ |

| $C_{2h}$ | E | $C_2$ | i | $\sigma_h$ | $h = 4$ | | |
|---|---|---|---|---|---|---|---|
| $A_g$ | 1 | 1 | 1 | 1 | $R_z$ | $x^2, y^2, z^2, xy$ |
| $B_g$ | 1 | $-1$ | 1 | $-1$ | $R_x, R_y$ | $xz, yz$ |
| $A_u$ | 1 | 1 | $-1$ | $-1$ | $z$ | |
| $B_u$ | 1 | $-1$ | $-1$ | 1 | $x, y$ | |

| $C_{3v}$ | E | $2C_3$ | $3\sigma_v$ | $h = 6$ | | |
|---|---|---|---|---|---|---|
| $A_1$ | 1 | 1 | 1 | $z$ | $x^2 + y^2, z^2$ |
| $A_2$ | 1 | 1 | $-1$ | $R_z$ | |
| $E$ | 2 | $-1$ | 0 | $(x, y), (R_x, R_y)$ | $(x^2 - y^2, xy), (xz, yz)$ |

| $C_{4v}$ | E | $2C_4$ | $C_2$ | $2\sigma_v$ | $2\sigma_d$ | $h = 8$ | | |
|---|---|---|---|---|---|---|---|---|
| $A_1$ | 1 | 1 | 1 | 1 | 1 | $z$ | $x^2 + y^2, z^2$ |
| $A_2$ | 1 | 1 | 1 | $-1$ | $-1$ | $R_z$ | |
| $B_1$ | 1 | $-1$ | 1 | 1 | $-1$ | | $x^2 - y^2$ |
| $B_2$ | 1 | $-1$ | 1 | $-1$ | 1 | | $xy$ |
| $E$ | 2 | 0 | $-2$ | 0 | 0 | $(x, y), (R_x, R_y)$ | $(xz, yz)$ |

| $D_{2h}$ | E | $C_2(z)$ | $C_2(y)$ | $C_2(x)$ | i | $\sigma(xy)$ | $\sigma(xz)$ | $\sigma(yz)$ | $h = 8$ | | |
|---|---|---|---|---|---|---|---|---|---|---|---|
| $A_g$ | 1 | 1 | 1 | 1 | 1 | 1 | 1 | 1 | | $x^2, y^2, z^2$ |
| $B_{1g}$ | 1 | 1 | $-1$ | $-1$ | 1 | 1 | $-1$ | $-1$ | $R_z$ | $xy$ |
| $B_{2g}$ | 1 | $-1$ | 1 | $-1$ | 1 | $-1$ | 1 | $-1$ | $R_y$ | $xz$ |
| $B_{3g}$ | 1 | $-1$ | $-1$ | 1 | 1 | $-1$ | $-1$ | 1 | $R_x$ | $yz$ |
| $A_u$ | 1 | 1 | 1 | 1 | $-1$ | $-1$ | $-1$ | $-1$ | | |
| $B_{1u}$ | 1 | 1 | $-1$ | $-1$ | $-1$ | $-1$ | 1 | 1 | $z$ | |
| $B_{2u}$ | 1 | $-1$ | 1 | $-1$ | $-1$ | 1 | $-1$ | 1 | $y$ | |
| $B_{3u}$ | 1 | $-1$ | $-1$ | 1 | $-1$ | 1 | 1 | $-1$ | $x$ | |

| $D_{3h}$ | E | $2C_3$ | $3C_2$ | $\sigma_h$ | $2S_3$ | $3\sigma_v$ | $h = 12$ | | |
|---|---|---|---|---|---|---|---|---|---|
| $A_1'$ | 1 | 1 | 1 | 1 | 1 | 1 | | $x^2 + y^2, z^2$ |
| $A_2'$ | 1 | 1 | $-1$ | 1 | 1 | $-1$ | $R_z$ | |
| $E'$ | 2 | $-1$ | 0 | 2 | $-1$ | 0 | $(x, y)$ | $(x^2 - y^2, xy)$ |
| $A_1''$ | 1 | 1 | 1 | $-1$ | $-1$ | $-1$ | | |
| $A_2''$ | 1 | 1 | $-1$ | $-1$ | $-1$ | 1 | $z$ | |
| $E''$ | 2 | $-1$ | 0 | $-2$ | 1 | 0 | $(R_x, R_y)$ | $(xz, yz)$ |

| $D_{4h}$ | E | $2C_4$ | $C_2$ | $2C_2'$ | $2C_2''$ | i | $2S_4$ | $\sigma_h$ | $2\sigma_v$ | $2\sigma_d$ | $h = 16$ | |
|---|---|---|---|---|---|---|---|---|---|---|---|---|
| $A_{1g}$ | 1 | 1 | 1 | 1 | 1 | 1 | 1 | 1 | 1 | 1 | | $x^2 + y^2, z^2$ |
| $A_{2g}$ | 1 | 1 | 1 | −1 | −1 | 1 | 1 | 1 | −1 | −1 | $R_z$ | |
| $B_{1g}$ | 1 | −1 | 1 | 1 | −1 | 1 | −1 | 1 | 1 | −1 | | $x^2 - y^2$ |
| $B_{2g}$ | 1 | −1 | 1 | −1 | 1 | 1 | −1 | 1 | −1 | 1 | | $xy$ |
| $E_g$ | 2 | 0 | −2 | 0 | 0 | 2 | 0 | −2 | 0 | 0 | $(R_x, R_y)$ | $(xz, yz)$ |
| $A_{1u}$ | 1 | 1 | 1 | 1 | 1 | −1 | −1 | −1 | −1 | −1 | | |
| $A_{2u}$ | 1 | 1 | 1 | −1 | −1 | −1 | −1 | −1 | 1 | 1 | $z$ | |
| $B_{1u}$ | 1 | −1 | 1 | 1 | −1 | −1 | 1 | −1 | −1 | 1 | | |
| $B_{2u}$ | 1 | −1 | 1 | −1 | 1 | −1 | 1 | −1 | 1 | −1 | | |
| $E_u$ | 2 | 0 | −2 | 0 | 0 | −2 | 0 | 2 | 0 | 0 | $(x, y)$ | |

| $D_{5h}$ | E | $2C_5$ | $2C_5^2$ | $5C_2$ | $\sigma_h$ | $2S_5$ | $2S_5^3$ | $5\sigma_v$ | $h = 20$ | |
|---|---|---|---|---|---|---|---|---|---|---|
| $A_1'$ | 1 | 1 | 1 | 1 | 1 | 1 | 1 | 1 | | $x^2 + y^2, z^2$ |
| $A_2'$ | 1 | 1 | 1 | −1 | 1 | 1 | 1 | −1 | $R_z$ | |
| $E_1'$ | 2 | $2\cos 72°$ | $2\cos 144°$ | 0 | 2 | $2\cos 72°$ | $2\cos 144°$ | 0 | $(x, y)$ | |
| $E_2'$ | 2 | $2\cos 144°$ | $2\cos 72°$ | 0 | 2 | $2\cos 144°$ | $2\cos 72°$ | 0 | | $(x^2 - y^2, xy)$ |
| $A_1''$ | 1 | 1 | 1 | 1 | −1 | −1 | −1 | −1 | | |
| $A_2''$ | 1 | 1 | 1 | −1 | −1 | −1 | −1 | 1 | $z$ | |
| $E_1''$ | 2 | $2\cos 72°$ | $2\cos 144°$ | 0 | −2 | $-2\cos 72°$ | $-2\cos 144°$ | 0 | $(R_x, R_y)$ | $(xz, yz)$ |
| $E_2''$ | 2 | $2\cos 144°$ | $2\cos 72°$ | 0 | −2 | $-2\cos 144°$ | $-2\cos 72°$ | 0 | | |

| $D_{6h}$ | E | $2C_6$ | $2C_3$ | $C_2$ | $3C_2'$ | $3C_2''$ | i | $2S_3$ | $2S_6$ | $\sigma_h$ | $3\sigma_d$ | $3\sigma_v$ | $h = 24$ | |
|---|---|---|---|---|---|---|---|---|---|---|---|---|---|---|
| $A_{1g}$ | 1 | 1 | 1 | 1 | 1 | 1 | 1 | 1 | 1 | 1 | 1 | 1 | | $x^2 + y^2, z^2$ |
| $A_{2g}$ | 1 | 1 | 1 | 1 | −1 | −1 | 1 | 1 | 1 | 1 | −1 | −1 | $R_z$ | |
| $B_{1g}$ | 1 | −1 | 1 | −1 | 1 | −1 | 1 | −1 | 1 | −1 | 1 | −1 | | |
| $B_{2g}$ | 1 | −1 | 1 | −1 | −1 | 1 | 1 | −1 | 1 | −1 | −1 | 1 | | |
| $E_{1g}$ | 2 | 1 | −1 | −2 | 0 | 0 | 2 | 1 | −1 | −2 | 0 | 0 | $(R_x, R_y)$ | $(xz, yz)$ |
| $E_{2g}$ | 2 | −1 | −1 | 2 | 0 | 0 | 2 | −1 | −1 | 2 | 0 | 0 | | $(x^2 - y^2, xy)$ |
| $A_{1u}$ | 1 | 1 | 1 | 1 | 1 | 1 | −1 | −1 | −1 | −1 | −1 | −1 | | |
| $A_{2u}$ | 1 | 1 | 1 | 1 | −1 | −1 | −1 | −1 | −1 | −1 | 1 | 1 | $z$ | |
| $B_{1u}$ | 1 | −1 | 1 | −1 | 1 | −1 | −1 | 1 | −1 | 1 | −1 | 1 | | |
| $B_{2u}$ | 1 | −1 | 1 | −1 | −1 | 1 | −1 | 1 | −1 | 1 | 1 | −1 | | |
| $E_{1u}$ | 2 | 1 | −1 | −2 | 0 | 0 | −2 | −1 | 1 | 2 | 0 | 0 | $(x, y)$ | |
| $E_{2u}$ | 2 | −1 | −1 | 2 | 0 | 0 | −2 | 1 | 1 | −2 | 0 | 0 | | |

| $D_{2d}$ | E | $2S_4$ | $C_2$ | $2C_2'$ | $2\sigma_d$ | $h = 8$ | |
|---|---|---|---|---|---|---|---|
| $A_1$ | 1 | 1 | 1 | 1 | 1 | | $x^2 + y^2, z^2$ |
| $A_2$ | 1 | 1 | 1 | −1 | −1 | $R_z$ | |
| $B_1$ | 1 | −1 | 1 | 1 | −1 | | $x^2 - y^2$ |
| $B_2$ | 1 | −1 | 1 | −1 | 1 | $z$ | $xy$ |
| $E$ | 2 | 0 | −2 | 0 | 0 | $(x, y), (R_x, R_y)$ | $(xz, yz)$ |

| $D_{3d}$ | E | $2C_3$ | $3C_2$ | i | $2S_6$ | $3\sigma_d$ | $h = 12$ | |
|---|---|---|---|---|---|---|---|---|
| $A_{1g}$ | 1 | 1 | 1 | 1 | 1 | 1 | | $x^2 + y^2, z^2$ |
| $A_{2g}$ | 1 | 1 | −1 | 1 | 1 | −1 | $R_z$ | |
| $E_g$ | 2 | −1 | 0 | 2 | −1 | 0 | $(R_x, R_y)$ | $(x^2 - y^2, xy), (xz, yz)$ |
| $A_{1u}$ | 1 | 1 | 1 | −1 | −1 | −1 | | |
| $A_{2u}$ | 1 | 1 | −1 | −1 | −1 | 1 | $z$ | |
| $E_u$ | 2 | −1 | 0 | −2 | 1 | 0 | $(x, y)$ | |

| $D_{4d}$ | E | $2S_8$ | $2C_4$ | $2S_8^3$ | $C_2$ | $4C_2'$ | $4\sigma_d$ | $h = 16$ | |
|---|---|---|---|---|---|---|---|---|---|
| $A_1$ | 1 | 1 | 1 | 1 | 1 | 1 | 1 | | $x^2 + y^2, z^2$ |
| $A_2$ | 1 | 1 | 1 | 1 | 1 | −1 | −1 | $R_z$ | |
| $B_1$ | 1 | −1 | 1 | −1 | 1 | 1 | −1 | | |
| $B_2$ | 1 | −1 | 1 | −1 | 1 | −1 | 1 | $z$ | |
| $E_1$ | 2 | $\sqrt{2}$ | 0 | $-\sqrt{2}$ | −2 | 0 | 0 | $(x, y)$ | |
| $E_2$ | 2 | 0 | −2 | 0 | 2 | 0 | 0 | | $(x^2 - y^2, xy)$ |
| $E_3$ | 2 | $-\sqrt{2}$ | 0 | $\sqrt{2}$ | −2 | 0 | 0 | $(R_x, R_y)$ | $(xz, yz)$ |

| $D_{5d}$ | E | $2C_5$ | $2C_5^2$ | $5C_2$ | i | $2S_{10}^3$ | $2S_{10}$ | $5\sigma_d$ | $h = 20$ | |
|---|---|---|---|---|---|---|---|---|---|---|
| $A_{1g}$ | 1 | 1 | 1 | 1 | 1 | 1 | 1 | 1 | | $x^2 + y^2, z^2$ |
| $A_{2g}$ | 1 | 1 | 1 | −1 | 1 | 1 | 1 | −1 | $R_z$ | |
| $E_{1g}$ | 2 | $2\cos 72°$ | $2\cos 144°$ | 0 | 2 | $2\cos 72°$ | $2\cos 144°$ | 0 | $(R_x, R_y)$ | $(xz, yz)$ |
| $E_{2g}$ | 2 | $2\cos 144°$ | $2\cos 72°$ | 0 | 2 | $2\cos 144°$ | $2\cos 72°$ | 0 | | $(x^2 - y^2, xy)$ |
| $A_{1u}$ | 1 | 1 | 1 | 1 | −1 | −1 | −1 | −1 | | |
| $A_{2u}$ | 1 | 1 | 1 | −1 | −1 | −1 | −1 | 1 | $z$ | |
| $E_{1u}$ | 2 | $2\cos 72°$ | $2\cos 144°$ | 0 | −2 | $-2\cos 72°$ | $-2\cos 144°$ | 0 | $(x, y)$ | |
| $E_{2u}$ | 2 | $2\cos 144°$ | $2\cos 72°$ | 0 | −2 | $-2\cos 144°$ | $-2\cos 72°$ | 0 | | |

| $D_{6d}$ | E | $2S_{12}$ | $2C_6$ | $2S_4$ | $2C_3$ | $2S_{12}^5$ | $C_2$ | $6C_2'$ | $6\sigma_d$ | $h = 24$ | |
|---|---|---|---|---|---|---|---|---|---|---|---|
| $A_1$ | 1 | 1 | 1 | 1 | 1 | 1 | 1 | 1 | 1 | | $x^2 + y^2, z^2$ |
| $A_2$ | 1 | 1 | 1 | 1 | 1 | 1 | 1 | -1 | -1 | $R_z$ | |
| $B_1$ | 1 | -1 | 1 | -1 | 1 | -1 | 1 | 1 | -1 | | |
| $B_2$ | 1 | -1 | 1 | -1 | 1 | -1 | 1 | -1 | 1 | $z$ | |
| $E_1$ | 2 | $\sqrt3$ | 1 | 0 | -1 | $-\sqrt3$ | -2 | 0 | 0 | $(x, y)$ | |
| $E_2$ | 2 | 1 | -1 | -2 | -1 | 1 | 2 | 0 | 0 | | $(x^2 - y^2, xy)$ |
| $E_3$ | 2 | 0 | -2 | 0 | 2 | 0 | -2 | 0 | 0 | | |
| $E_4$ | 2 | -1 | -1 | 2 | -1 | -1 | 2 | 0 | 0 | | |
| $E_5$ | 2 | $-\sqrt3$ | 1 | 0 | -1 | $\sqrt3$ | -2 | 0 | 0 | $(R_x, R_y)$ | $(xz, yz)$ |

## Cubic point groups

| $T_d$ | E | $8C_3$ | $3C_2$ | $6S_4$ | $6\sigma_d$ | $h = 24$ | |
|---|---|---|---|---|---|---|---|
| $A_1$ | 1 | 1 | 1 | 1 | 1 | | $x^2 + y^2 + z^2$ |
| $A_2$ | 1 | 1 | 1 | -1 | -1 | | |
| $E$ | 2 | -1 | 2 | 0 | 0 | | $(2z^2 - x^2 - y^2, x^2 - y^2)$ |
| $T_1$ | 3 | 0 | -1 | 1 | -1 | $(R_x, R_y, R_z)$ | |
| $T_2$ | 3 | 0 | -1 | -1 | 1 | $(x, y, z)$ | $(xy, xz, yz)$ |

| $O_h$ | E | $8C_3$ | $6C_2$ | $6C_4$ | $3C_2(= C_4^2)$ | i | $6S_4$ | $8S_6$ | $3\sigma_h$ | $6\sigma_d$ | $h = 48$ | |
|---|---|---|---|---|---|---|---|---|---|---|---|---|
| $A_{1g}$ | 1 | 1 | 1 | 1 | 1 | 1 | 1 | 1 | 1 | 1 | | $x^2 + y^2 + z^2$ |
| $A_{2g}$ | 1 | 1 | -1 | -1 | 1 | 1 | -1 | 1 | 1 | -1 | | |
| $E_g$ | 2 | -1 | 0 | 0 | 2 | 2 | 0 | -1 | 2 | 0 | | $(2z^2 - x^2 - y^2, x^2 - y^2)$ |
| $T_{1g}$ | 3 | 0 | -1 | 1 | -1 | 3 | 1 | 0 | -1 | -1 | $(R_x, R_y, R_z)$ | |
| $T_{2g}$ | 3 | 0 | 1 | -1 | -1 | 3 | -1 | 0 | -1 | 1 | | $(xz, yz, xy)$ |
| $A_{1u}$ | 1 | 1 | 1 | 1 | 1 | -1 | -1 | -1 | -1 | -1 | | |
| $A_{2u}$ | 1 | 1 | -1 | -1 | 1 | -1 | 1 | -1 | -1 | 1 | | |
| $E_u$ | 2 | -1 | 0 | 0 | 2 | -2 | 0 | 1 | -2 | 0 | | |
| $T_{1u}$ | 3 | 0 | -1 | 1 | -1 | -3 | -1 | 0 | 1 | 1 | $(x, y, z)$ | |
| $T_{2u}$ | 3 | 0 | 1 | -1 | -1 | -3 | 1 | 0 | 1 | -1 | | |

## Linear point groups

| $C_{\infty v}$ | E | $2C_\infty^\phi$ | $\cdots$ | $\infty\sigma_v$ | | |
|---|---|---|---|---|---|---|
| $A_1 \equiv \sum^+$ | 1 | 1 | $\cdots$ | 1 | $z$ | $x^2+y^2, z^2$ |
| $A_2 \equiv \sum^-$ | 1 | 1 | $\cdots$ | $-1$ | $R_z$ | |
| $E_1 \equiv \prod$ | 2 | $2\cos\Phi$ | $\cdots$ | 0 | $(x,y), (R_x, R_y)$ | $(xz, yz)$ |
| $E_2 \equiv \Delta$ | 2 | $2\cos2\Phi$ | $\cdots$ | 0 | | $(x^2-y^2, xy)$ |
| $E_3 \equiv \Phi$ | 2 | $2\cos3\Phi$ | $\cdots$ | 0 | | |
| $\cdots$ | $\cdots$ | $\cdots$ | | $\cdots$ | $\cdots$ | |

| $D_{\infty h}$ | E | $2C_\infty^\Phi$ | $\cdots$ | $\infty\sigma_v$ | $i$ | $2S_\infty^\Phi$ | $\cdots$ | $\infty C_2$ | | |
|---|---|---|---|---|---|---|---|---|---|---|
| $\sum_g^+$ | 1 | 1 | $\cdots$ | 1 | 1 | 1 | $\cdots$ | 1 | | $x^2+y^2, z^2$ |
| $\sum_g^-$ | 1 | 1 | $\cdots$ | $-1$ | 1 | 1 | $\cdots$ | $-1$ | $R_z$ | |
| $\prod_g$ | 2 | $2\cos\Phi$ | $\cdots$ | 0 | 2 | $-2\cos\Phi$ | $\cdots$ | 0 | $(R_x, R_y)$ | $(xz, yz)$ |
| $\Delta_g$ | 2 | $2\cos2\Phi$ | $\cdots$ | 0 | 2 | $2\cos2\Phi$ | $\cdots$ | 0 | | $(x^2-y^2, xy)$ |
| $\cdots$ | $\cdots$ | $\cdots$ | | $\cdots$ | $\cdots$ | $\cdots$ | $\cdots$ | | | |
| $\sum_u^+$ | 1 | 1 | $\cdots$ | 1 | $-1$ | $-1$ | $\cdots$ | $-1$ | $z$ | |
| $\sum_u^-$ | 1 | 1 | $\cdots$ | $-1$ | $-1$ | $-1$ | $\cdots$ | 1 | | |
| $\prod_u$ | 2 | $2\cos\Phi$ | $\cdots$ | 0 | $-2$ | $2\cos\Phi$ | $\cdots$ | 0 | $(x,y)$ | |
| $\Delta_u$ | 2 | $2\cos2\Phi$ | $\cdots$ | 0 | $-2$ | $-2\cos2\Phi$ | $\cdots$ | 0 | | |
| $\cdots$ | $\cdots$ | $\cdots$ | | $\cdots$ | $\cdots$ | $\cdots$ | $\cdots$ | | | |

# Appendix III

**Bibliography**

**General textbooks**

D. F. Shriver and P. W. Atkins, *Inorganic Chemistry*, 3rd edn, Oxford University Press, Oxford, 1999.

D. M. P. Mingos, *Essential Trends in Inorganic Chemistry*, Oxford University Press, Oxford, 1998.

**Symmetry**

R. L. Carter, *Molecular Symmetry and Group Theory*, John Wiley and Sons, New York, 1998.

P. W. Atkins, M. S. Child and C. S. G. Phillips, *Tables for Group Theory*, Oxford University Press, Oxford, 1970.

**Spectroscopy**

A. K. Brisdon, *Inorganic Spectroscopic Methods*, Oxford Chemistry Primer No. 62, Oxford, 1998.

**Bonding**

M. J. Winter, *Chemical Bonding*, Oxford Chemistry Primer No. 15, Oxford, 1994.

# Index

**DATE DUE**

OhioLINK
AUG 0 1 REC'D
MAY 1 7 2004
MAY 1 7 2004

MAY 1 6 2009
FEB 2 7 2005
AUG 2 2 2005
JUL 1 2 2005
DEC 2 3 2005
DEC 2 1 2005
MAY 1 6 2006
MAY 1 1 2006
DEC 1 7 2012

GAYLORD                                    PRINTED IN U.S.A.

SCI QD 461 .O365 2001

Ogden, J. S.

Introduction to molecular
  symmetry